The Practical Approach in Chemistry Series

SERIES EDITORS

L. M. Harwood
Department of Chemistry
University of Reading

C. J. Moody
Department of Chemistry
University of Exeter

The Practical Approach in Chemistry Series

Organocopper reagents
Edited by Richard J. K. Taylor

Macrocycle synthesis
Edited by David Parker

Preparation of alkenes
Edited by Jonathan M. J. Williams

High-pressure techniques in chemistry and physics
Edited by Wilfried B. Holzapfel and Neil S. Isaacs

Transition metals in organic synthesis
Edited by Susan E. Gibson (née Thomas)

Matrix-isolation techniques
Ian R. Dunkin

Matrix-Isolation Techniques

A Practical Approach

IAN R. DUNKIN

Department of Pure and Applied Chemistry
University of Strathclyde, Glasgow, UK

OXFORD   NEW YORK   TOKYO
OXFORD   UNIVERSITY   PRESS
1998

Oxford University Press, Great Clarendon Street, Oxford OX2 6DP

Oxford   New York
Athens   Auckland   Bangkok   Bogota   Bombay   Buenos Aires
Calcutta   Cape Town   Dar es Salaam   Delhi   Florence   Hong Kong
Istanbul   Karachi   Kuala Lumpur   Madras   Madrid   Melbourne
Mexico City   Nairobi   Paris   Singapore   Taipei   Tokyo   Toronto   Warsaw

and associated companies in
Berlin   Ibadan

Oxford is a trade mark of Oxford University Press

Published in the United States
by Oxford University Press Inc., New York

© Ian R. Dunkin, 1998

A catalogue record for this book is available from the British Library

Library of Congress Cataloging-in-Publication Data
Dunkin, Ian R.
Matrix-isolation techniques : a practical approach / Ian R. Dunkin.
(The practical approach in chemistry series)
Includes bibliographical references and index.
1. Matrix isolation spectroscopy. I. Title. II. Series.
QD96.M33D86   1998   543'.0858—dc21   97-32141

ISBN 0 19 855863 5 (Hbk)

Typeset by Footnote Graphics, Warminster, Wilts
Printed in Great Britain by
Bookcraft (Bath) Ltd
Midsomer Norton, Avon

Preface

Research in matrix isolation has provided me with great pleasure and excitement for many years, and yet my introduction to the technique was quite accidental. An organic chemist by training, I was a postdoctoral member of Leo Paquette's group at Ohio State University when, in late 1973, the first reports of the matrix IR spectrum of cyclobutadiene appeared. I remember reading these with amazement, but with no suspicion that, by the end of the same year, I would be actively involved in matrix isolation. In September 1973 I returned to the UK. Industrial posts in chemistry were thin on the ground and a further postdoctoral research post, which had been arranged, fell through at the last moment. For several weeks, therefore, I found myself unemployed and scanning the appointments sections in *New Scientist* and *Chemistry in Britain*. I eventually saw an advertisement for an organic chemist to join Jim Turner, Martyn Poliakoff, and Mark Baird at Newcastle University, to carry out matrix studies of carbenes. Still preferring a career in industry, I nevertheless applied for this post, and was soon in Newcastle learning how to carry out matrix experiments. I was therefore fortunate to get involved in the application of matrix isolation to organic chemistry at an early stage in the evolution of the subject. This turned out to be one of those all-too-rare pieces of career serendipity.

The group at Newcastle at that time provided a tremendously stimulating environment, and I soon gave up the idea of an industrial career, at least for the time being. As the months passed, matrix isolation research seemed more and more exciting, and it became clear that there was a great deal of matrix organic chemistry to be explored. I eventually moved to a permanent academic post in Scotland at Strathclyde University, where I set up my own matrix-isolation laboratory, based on know-how acquired from Jim Turner and Martyn Poliakoff, who both moved to Nottingham University at about the same time. The success of this venture depended heavily on the intensely practical training that I had received at Newcastle. Martyn, in particular, has an unrivalled genius for turning mundane objects, such as jam-jar lids, into useful scientific gadgets; and his ceaseless desire to innovate has rubbed off on me at least to some extent.

The preceding paragraphs of personal history show that I came to matrix isolation from a background in conventional preparative and mechanistic organic chemistry, and that the techniques needed to pursue matrix research can be learnt easily enough, given the right tuition. This book is intended to provide an introduction to matrix isolation for those with little or no previous experience of the subject. It deals with the general scope and limitations of matrix isolation, the best methods of preparing matrices, and a selection of

Preface

techniques for generating reactive species in matrices. It should provide the necessary basic knowledge for researchers joining or intending to collaborate with a matrix group; while for those with sufficient ambition (and funding), it describes how to set up a matrix-isolation laboratory from scratch.

Of course, this book reflects my own background and bias towards organic chemistry, but if this encourages other organic chemists to consider matrix isolation as an accessible mechanistic tool, it will not be too great a defect. In any case, I hope the book will help to stimulate continued interest in all areas of matrix-isolation research amongst a widening group of workers.

I. R. D.

Glasgow
December 1997

Contents

Contents

Abbreviations

b.p.	boiling point
DMF	dimethylformamide
DNA	deoxyribonucleic acid
EPA	ether/isopentane/ethanol mixture
ESR	electron spin resonance
EXAFS	extended X-ray absorption fine-structure spectroscopy
FTIR	Fourier-transform infrared spectroscopy
GC	gas chromatography
HPLC	high-pressure liquid chromatography
i.d.	internal diameter
IR	infrared
MCD	magnetic circular dichroism
m.p.	melting point
MS	mass-spectrometry
NMR	nuclear magnetic resonance
o.d.	outer diameter
ppm	parts per million
PTFE	poly(tetrafluoroethene)
PVC	poly(vinyl chloride)
UV	ultraviolet

Acknowledgements

Writing this book would have been impossible without the stimuli that I have received from many others involved in matrix-isolation research. Special thanks are due to Jim Turner and Martyn Poliakoff, who introduced me to matrix isolation, taught me most of the necessary practical skills, and then encouraged me to take an academic post and continue my research in this area. I should also like to thank Lester Andrews, with whom I spent an enjoyable and fruitful period of sabbatical leave, investigating organic radical cations in matrices, and the many other friends whom I have made through matrix-isolation conferences, sharing knowledge and alcohol in roughly equal measures.

Next, I must acknowledge the contribution of my Research Students and Postdoctoral Assistants, who have continued to produce new results, keeping me excited over more than 20 years. They have even allowed me to use the equipment occasionally.

Drs Rob Withnall and Charles Gordon were kind enough to read most of the typescript, and suggested numerous improvements. I thank them both, while emphasizing that all remaining defects are the authors' sole responsibility.

Finally, I dedicate this book to my wife, Janet, and my parents. They have encouraged me throughout the period during which the book was written, which was rather longer than it should have been, and no doubt prevented even worse delays by regular and embarrassing enquiries as to its progress. I hope they will survive the shock of discovering that it is at last complete.

1

Matrix isolation—an introduction to the technique

In the most general sense, matrix isolation comprises a range of experimental techniques in which *guest* molecules or atoms are trapped in rigid *host* materials. The trapped species are prevented from diffusing, and cannot therefore undergo any bimolecular reactions, except with the host (Fig. 1.1). The host material can be a crystalline solid, a polymer, or a glass formed by freezing a liquid or solidifying a gas. The term *matrix isolation* is, however, most commonly used in a narrower sense: to refer to the technique of trapping molecules or atoms in solidified inert (or occasionally reactive) gases—a technique requiring very low temperatures.

This book is about matrix isolation in the latter, more restricted sense. It deals with the practicalities of carrying out matrix-isolation experiments, describing in detail the design and construction of the necessary equipment and the best methods of operating it to achieve successful results. The detailed practical advice found throughout the book will, it is hoped, be helpful to those who are planning some experiments in collaboration with an existing matrix-isolation group, to those who are joining a matrix group but have no previous experience of matrix isolation, and finally to those who are contemplating setting up a matrix-isolation laboratory from scratch.

Before we get to these practical matters, however, and in order to under-

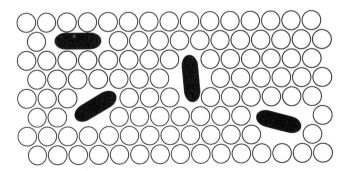

Fig. 1.1 Matrix isolation. The rigid host lattice (shown as open circles) isolates molecules of a reactive species from each other and prevents bimolecular reaction.

stand the advantages and limitations of matrix isolation, it would be wise to consider how and why the technique was developed and to take an overview of its scope.

1. A brief history of matrix isolation

1.1 Early days

In the latter part of the nineteenth century, simple methods of liquefying gases were discovered, which provided chemists and physicists with a convenient means of carrying out studies at low temperatures. It was soon observed in several laboratories that the phosphorescence of certain organic molecules was enhanced at low temperatures. Many organic compounds which gave no phosphorescence at room temperature emitted light when cooled to 77 K (the boiling point of liquid nitrogen). In these early studies, the materials of interest were simply immersed in the coolant (liquid N_2 or liquid air), and visible emission was stimulated by irradiation with X-rays or UV light. For 'cutting-edge' research, the apparatus was exceedingly simple (Fig. 1.2). Although not appreciated until much later, the enhancement of phosphorescence was due, at least in part, to suppressed diffusion of triplet quenchers such as molecular oxygen. As a result, the lifetimes of the triplet excited states responsible for the emission were prolonged. These experiments may thus have been the first, albeit unwitting, application of the matrix-isolation principle.

For the first three decades of the twentieth century, experiments in low-temperature photochemistry and photophysics were confined to emission studies, mainly because of the poor optical quality of solutions frozen and cooled to 77 K. Most solvents become crazed, or cloudy in these conditions, and are thus highly light-scattering, especially at short wavelengths. With such optically poor samples, emission can be studied but not absorption.

In the 1930s, searches were made, particularly by Gilbert Lewis and his

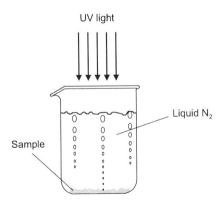

Fig. 1.2 Viewing enhanced phosphorescence at 77 K.

group, to find solvent mixtures which froze to clear glasses with good optical properties. One of the best turned out to be EPA, a mixture of ether, isopentane and alcohol (ethanol), typically in the ratio 5:5:2. Provided the proportion of ethanol is not increased excessively, EPA forms glasses at 77 K which are clear and transparent throughout the UV–visible region. This solvent combination is still widely used.

With the development of EPA glasses, it became possible to generate reactive species in frozen solutions and record their UV–visible absorption spectra. For example, Lewis[1] showed that UV-irradiation of 1,4-bis(dimethylamino) benzene in EPA gave the same radical cation (Wurster's blue) that had previously been generated by chemical oxidation[2] (Scheme 1.1 and Fig. 1.3). In the low-temperature photoprocess, an electron must have escaped and become solvated by one of the host molecules. Similar radical cations were observed after photoionization of other aromatic amines.

The UV-irradiation and UV–visible spectroscopy were carried out with the sample contained in a quartz ampoule, which was cooled by suspending it above liquid air in an unsilvered quartz Dewar vessel (Fig. 1.4). Cooling by

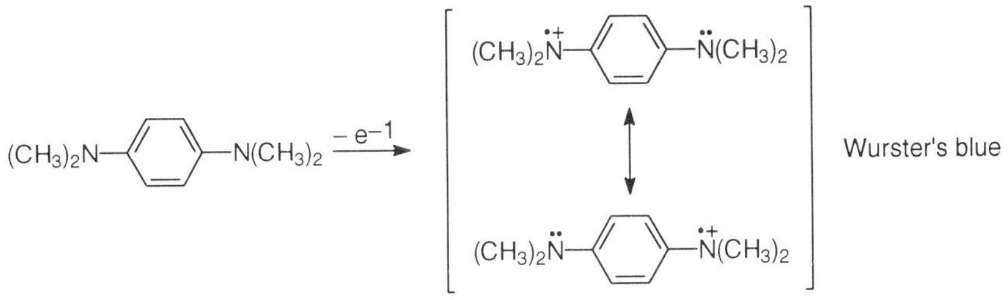

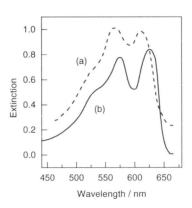

Scheme 1.1

Fig. 1.3 Visible absorption spectra of Wurster's blue generated (a) by chemical oxidation in aqueous methanol solutions (ref. 2) and (b) by photoionization in EPA at 90 K (ref. 1). The two spectra are remarkably similar, allowing for the differences of medium and temperature. (Adapted with permission from ref. 1. © 1942, American Chemical Society.)

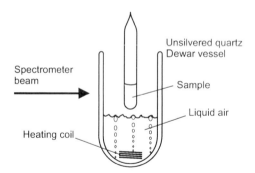

Fig. 1.4 A simple cryostat for absorption spectroscopy in EPA glasses at temperatures down to 90 K. Heating the liquid air increases its rate of boiling and thus reduces the sample temperature.

immersion of the ampoule in the liquid air did not work so well, because the spectrometer beam then had to pass through the boiling coolant with its numerous bubbles. The result was excessive light scattering, making spectra difficult or impossible to record. By suspending the sample in the cold gas above the liquid coolant, this problem was avoided. The sample temperature could be controlled to some extent by varying the rate of boiling of the liquid air. A coil of heating wire was placed in the Dewar vessel, and running a current through this increased the boiling rate and thus reduced the sample temperature. Temperatures down to about 90 K could be maintained in this way (liquid air boils at about 80–90 K, depending on its oxygen content). A piece of equipment for maintaining low temperatures is known as a *cryostat*; Fig. 1.4 shows one of the earliest types of cryostat for absorption spectroscopy.

Reactive species could also be generated in EPA glasses by non-photolytic methods. For example, reductions of triarylmethyl halides, such as Ph_3CBr, with a silver–mercury amalgam were carried out in ampoules on a vacuum line (Fig. 1.5), yielding the corresponding triarylmethyl radicals.[3] The resulting solutions were quickly frozen in liquid air, transferred to a cryostat like the one shown in Fig. 1.4, and both fluorescence and absorption spectra of the radicals were obtained (Fig. 1.6).

Triarylmethyl radicals and radical cations such as Wurster's blue are relatively stable species, as reactive intermediates go. Clearly, the presence of ethanol in the solvent system must place limits on the reactivity of species which can be stabilized in EPA glasses, even allowing for the low temperatures employed. Nevertheless, studies utilizing EPA and other organic glasses continue to this day. More elaborate cryostats have been developed, and some surprisingly high-energy intermediates have been observed in this way. Of particular note are the extensive studies of free radicals made by George Porter's group[4] and those of Arnost Reiser and co-workers, who in the 1960s generated and characterized a range of triplet arylnitrenes in EPA glasses,

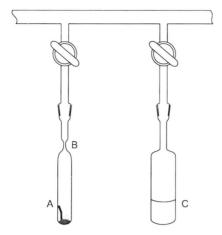

Fig. 1.5 Apparatus for the generation of triarylmethyl radicals. The triarylmethyl halide (e.g. Ph_3CBr) was placed in tube **A** together with a drop of mercury and a piece of silver wire; the tube was then evacuated and cooled by immersion in liquid air. Pre-dried solvent was distilled on to the resulting mixture from reservoir **C**, and **A** was then sealed off at **B** and allowed to warm. As soon as the mercury melted, the tube was shaken in the dark; the reaction usually took only a few minutes. Finally, the mixture was re-cooled, and spectra were recorded using a cryostat like that shown in Fig. 1.4.

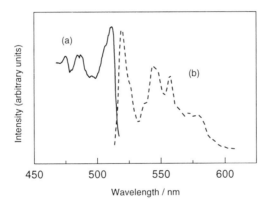

Fig. 1.6 Visible absorption (a) and fluorescence (b) spectra of triphenylmethyl radicals ($Ph_3C\cdot$) in EPA at 90 K. The radicals were generated as shown in Fig. 1.5. (Adapted with permission from ref. 3. © 1944, American Chemical Society.)

including phenylnitrene (PhN).[5-8] Despite almost complete reliance on UV–visible absorption spectroscopy, Reiser's original assignments have been confirmed by more recent low-temperature techniques.[9-11]

The few examples quoted above no doubt reflect the personal bias of the author, but a full history of early work with low-temperature glasses lies beyond the scope of this book. The interested reader can, however, find detailed accounts in books by Arnold Bass and Herbert Broida (1960)[12] and Beat Meyer (1971).[13] In their day, these were the 'Bibles' of low-temperature chemistry and physics. They are long since out of print, but can still be found in libraries. Surprisingly, Reiser's nitrene work was omitted from Meyer's otherwise comprehensive survey of low-temperature chemistry.

1.2 Solidified gas matrices

Low-temperature organic glasses have several disadvantages as trapping media. Most importantly, they are not chemically inert to very reactive species, such as metal atoms or carbenes, and they absorb strongly over large regions of the IR spectrum. Since much more structural information about trapped molecules can usually be derived from IR than from UV–visible spectroscopy, the latter is a severe problem.

It was George Pimentel who invented matrix isolation in solidified noble gases, which, as trapping media for reactive species, have the triple advantages of extreme chemical inertness, complete transparency throughout the normal IR–visible–UV regions of the spectrum, and a tendency to form clear glasses. The idea was first published in 1954 in a much quoted, single-page paper from Pimentel's group in the *Journal of Chemical Physics*.[14] The experimental advances reported in this paper were really quite small, because the lowest temperature initially available in Pimentel's laboratory was 66 K—cold enough to form xenon matrices, but too warm for neon, argon or krypton. Moreover, at this temperature even xenon formed rather 'soft' matrices, which failed to isolate species such as monomeric NO_2 satisfactorily, or prevent NH_3 or HN_3 from aggregating into H-bonded dimers and oligomers. Only the higher melting but much less inert matrix hosts, CO_2, CCl_4 and methylcyclohexane, successfully trapped any of these molecules as monomers. Nevertheless, the great idea was there and its potential was already appreciated; so 1954 is always regarded as the 'Anno Domini' of matrix isolation.

To form more rigid matrices from the noble gases, it was clearly necessary to attain lower temperatures in the cryostat, and for this there were only two reasonable choices of coolant: liquid hydrogen (b.p. 20 K) and liquid helium (b.p. 4.2 K). Neither is without hazard. Hydrogen is highly inflammable and forms explosive mixtures with oxygen. The condensation of solid oxygen or air into an open vessel of liquid hydrogen is especially dangerous. Liquid helium, on the other hand, has a very low heat of vaporization, which means that a small amount of heat input generates a large volume of gas. This can lead to an explosion if a Dewar vessel containing liquid helium loses its insulating vacuum or becomes plugged with solid air. Meyer's book[13] describes one such event, in which the explosion of a 25 litre Dewar containing 4 litres of liquid helium destroyed the entire contents of a laboratory (fortunately unoccupied at the time) and embedded 25 kg of metal from the top portion of the Dewar in the concrete ceiling, cutting off all electricity supplies to the laboratory. Today, the safe handling of liquid helium has become routine, especially with the widespread applications of superconducting magnets. In the first years of matrix-isolation studies, however, liquid hydrogen and liquid helium seem to have been regarded as more or less equally hazardous.

Figure 1.7 shows the main features of a liquid-hydrogen cryostat for matrix-isolation experiments. The matrices are deposited from the gas phase by condensation on to a cold spectroscopic window in a copper holder. The window holder is in thermal contact with the base of an inner vessel containing liquid hydrogen. The liquid-hydrogen vessel is separated by a vacuum from a surrounding annular vessel containing liquid nitrogen. An extension to the bottom of the liquid-nitrogen vessel forms a radiation shield around the cold window, thus minimizing heating effects from ambient radiation. The whole of this set-up is enclosed in a vacuum chamber fitted with a port for connection to the vacuum pumping system and external windows to allow passage of the beam from a spectrometer. A joint with O-ring seals enables the cold window to be rotated within the vacuum chamber.

Figure 1.8 is a simplified top view of the cold window in its vacuum chamber, and shows how matrix deposition and matrix spectroscopy can be accomplished. The cold window is rotated to face the inlet port in the vacuum chamber. A mixture of the host gas and a suitable guest is admitted via a control valve, and is deposited on one side of the cold window at about 20 K as

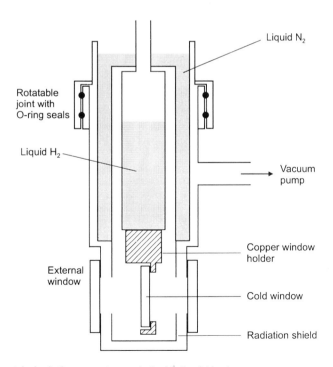

Fig. 1.7 A matrix-isolation cryostat cooled with liquid hydrogen.

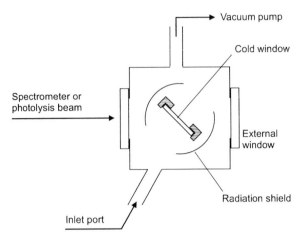

Fig. 1.8 Top view of a cryostat cold window and vacuum chamber.

a glassy matrix. Note that the matrix is insulated thermally from the environment by the surrounding vacuum, but is otherwise not enclosed in any sort of container, unlike a frozen solution in an ampoule. The temperature of the cold window must therefore be low enough not only to solidify the host gas but also to leave it with a negligible vapour pressure. Deposition of the gas mixture will also take place on the window holder (at 20 K) and, depending on the volatility of the host gas and guest species, may even occur on the radiation shield, which is at 77 K. With the usual sorts of absorption spectroscopy, however, only that portion of the matrix deposited on the transparent cold window can be examined. After deposition of the matrix, the cold window is rotated to line up with a pair of external windows. The matrix can then be analysed spectroscopically or irradiated with UV light.

The basic cryostat design was soon adapted for use with other types of spectroscopy. Samuel Foner and co-workers, for example, designed a cold cell for matrix ESR spectroscopy, and were thus able to obtain ESR spectra of matrix-isolated hydrogen, nitrogen, and alkali metal atoms, and small radicals such as NH_2 and ND_2.[15]

With cryostats working at 20 K, rigid matrices can be prepared with Ar, Kr, Xe, N_2, O_2, CH_4, CO, SF_6, and numerous other host materials, both reactive and inert. Missing from the list of accessible inert hosts are helium, which cannot be solidified, and neon, which requires a temperature of about 6 K or less to form sufficiently rigid matrices. A great deal of matrix-isolation research was carried out in the 1950s and 1960s on cryostats of this general design, but they were eventually superseded by commercial closed-cycle helium refrigerators, which are much more convenient to use. This work included studies of

- atomic species, mainly metal atoms
- diatomics (e.g. C_2, C_2^{2-}, CN, LiF, NF, OH, OF, ScO, Si_2, SiO and S_2)

- triatomics (e.g. Al_2O, BeF_2, $BeCl_2$, C_3, CCO, CCl_2, CNN, KrF_2, S_3, and $XeCl_2$)

- tetra-atomic and larger species (e.g. CH_3, CCl_3, S_4, SiF_3, and XeF_4).

The whole list is very long, but is covered fully in Meyer's book.[13]

Most of the research groups active in matrix isolation had, and still tend to have, direct links back to Pimentel's laboratory. Many of his former associates went on to set up matrix-isolation laboratories of their own, or took their expertise in matrix isolation to other established laboratories.

There has been a tendency for skills in matrix isolation to be handed down within the growing number of groups spreading out from George Pimentel's extended 'scientific family.' The author, for example, is one of Pimentel's numerous 'grandsons,' having learnt the technique at Newcastle University in the group of Jim Turner (now at Nottingham), who was one of Pimentel's postdoctoral associates. Despite the reasonably extensive list of books and review articles devoted to matrix isolation, it has still, in many instances, proved difficult for research workers to set up the technique in their own laboratories without extensive prior experience. There are many small but significant details of the design and operation of matrix-isolation equipment which are not mentioned in the published literature. It is certainly one of the main aims of this book to help fill this gap.

1.3 Organic chemists join in

Perhaps owing to the complexity of the equipment needed for matrix isolation, the nature of the experiments, and the history of its development, the technique tended at first to attract chemical physicists, physical chemists, and inorganic chemists. It was a milestone in the development of the subject, therefore, when in 1972–3 three organic groups published papers describing the IR spectrum of cyclobutadiene matrix isolated in argon or nitrogen.[16–18]

Cyclobutadiene was generated in the matrices by photolysis of the stable precursor α-pyrone (Scheme 1.2). The α-pyrone underwent a fast photo-equilibration with four rotational isomers of a ring-opened ketene and a much slower ring closure to a bicyclic β-lactone, which in turn eliminated CO_2 to give cyclobutadiene. Four IR bands were assigned to cyclobutadiene (3040, 1240, 650 and 570 cm^{-1}), and much debate ensued over whether the molecule was rectangular, as predicted theoretically, or square. The interpretation of the IR spectrum of this early example of a 'larger' reactive organic species in matrices was very instructive. One of the IR bands originally assigned to cyclobutadiene (at 650 cm^{-1}) turned out to belong to molecules of CO_2 which remained trapped alongside cyclobutadiene molecules in the matrix. The close association of the two species perturbed the CO_2 bending frequency from its normal value of 667 cm^{-1}. It took another five years, during which matrix-isolated cyclobutadiene was generated from a variety of precursors, before a

9

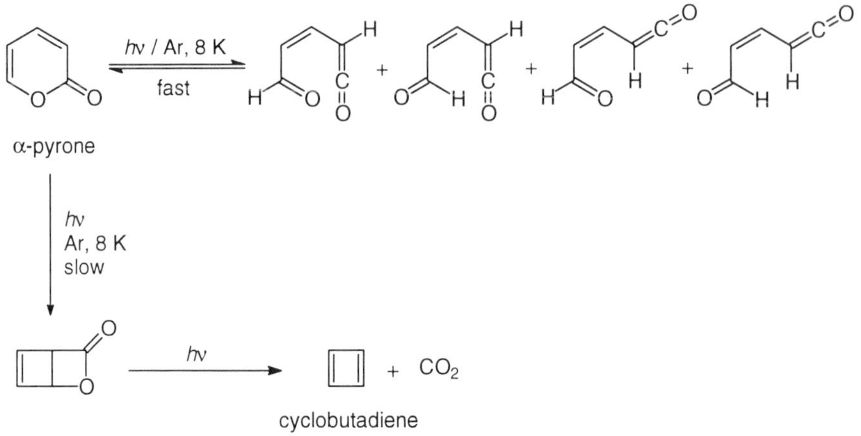

Scheme 1.2

definitive matrix IR spectrum for this species finally excluded the possibility of a square structure.[19]

Despite the initial difficulties over spectral interpretation, the observation of matrix isolated cyclobutadiene was a real advance, and the application of matrix isolation to problems in mechanistic organic chemistry caught the imagination of the organic chemical community. There followed matrix IR, UV–visible and ESR studies of unstable molecules such as arynes, carbenes, nitrenes, radical anions and numerous others. Organic matrix laboratories now exist all over the world; the author has contacts with organic matrix groups in five continents. The contribution of matrix isolation to organic chemistry in general was first reviewed in 1980,[20] with an updated review appearing in 1989.[21]

1.4 Matrix isolation today

Modern cryostats are commercially available and can be incorporated into a variety of experimental systems, some of which are described in Chapters 2 and 3. The practice of matrix isolation is certainly now widespread geographically, but there are still relatively few research groups with the necessary equipment. Therefore much collaboration exists between chemists with specific mechanistic problems and those groups with the right matrix set-up to tackle them. There is also a tendency for each matrix laboratory to have its own specializations in spectroscopic techniques or methods for generating reactive species, for example photolysis, X-irradiation or pyrolysis. Some forms of spectroscopy are very specialized and found in only a few laboratories, such as magnetic circular dichroism and matrix NMR; thus collaboration between matrix groups is also common.

More and more challenging molecules are being studied, such as binuclear metal complexes, highly photolabile carbonyl oxides, and DNA bases. The applications of matrix isolation are by no means confined to the investigation of reactive species. The formation of complexes between stable species, including hydrogen-bonding, has been studied in matrices since the earliest days, while more recently there have been many novel developments in matrix physics, for example with IR emission studies. Matrix isolation even finds application in chemical analysis. One company (Mattson) offers for sale a GC–MS–FTIR system, in which part of the eluate from the gas chromatography (GC) column can be analysed by mass spectrometry (MS) while the rest is deposited as a helical matrix on the edge a cold, rotating, gold-plated drum. After elution is complete, the individual components of the eluate can be analysed by matrix FTIR reflection spectroscopy on the cold drum, and correlations between the IR and mass spectra can be made.

Since 1977, there has been a series of biennial international conferences on matrix isolation. These are held in the summer and, at present, alternate between a Gordon research conference in New Hampshire and a European conference. Attendance at the conferences has remained at about 100–125 delegates, which is a significant proportion of the total number of practising matrix chemists and physicists. The programme has traditionally been multidisciplinary, with contributions from theoreticians and physicists at one end of the spectrum and organic chemists at the other. There still seems to be a lively interest in the development of new techniques and a broadening of the range of matrix experiments. It is to be hoped that the scope and limitations of matrix isolation will steadily become better understood by a wider section of the chemistry and physics community. Promoting such an understanding is one of the aims of this book.

George Pimentel died in 1989, sadly early at the age of 67 and while still very active. Until his death, and despite his heavy involvement at the highest levels of American science and science-policy making, he managed to attend the matrix conferences and take an interest in all the new participants and their research. Appreciations of George Pimentel appeared in many publications; particularly recommended to the reader is a biography in the *Journal of Physical Chemistry*, which was written by four of his very close associates and which contains a full bibliography.[22]

1.5 Books and reviews

Although there is no available space in a practical book of this type for extensive surveys of the literature of matrix isolation, the reader who wishes to find out more about specific areas of matrix-isolation research can at least be pointed in the right direction. For the earliest work, the books by Bass and Broida[12] and by Meyer[13] provide the best starting points for any literature search. Following these, books on matrix isolation have appeared at irregular intervals. These include books with an emphasis on metal atom and inorganic chemistry,[23,24] or vibrational spectroscopy,[25] as well as those which attempt a more or less general coverage.[26,27] All these provide some details of experimental techniques and most are well referenced.

Of particular usefulness is a bibliography of matrix-isolation spectroscopy, covering the literature from 1954 to 1985, which is arranged in a sensibly compact way, and indexed under author, chemical formula, and key words.[28] It is not clear whether this bibliography will be updated in the future, but its literature coverage already overlaps conveniently with the publication dates at which on-line searching can take over. The *Science Citation Index*, for example, can be accessed on-line in the UK through BIDS (Bath Information and Data Services), and provides literature coverage back to 1981. The title, abstract, and key-word fields of the database can be searched with a boolean formula such as:

(matrix_isolat*),(low_temp*+(matri*,spectro*)).

Together with any other appropriate search criteria, this gives a reasonably efficient route into specific sections or the whole of the more recent matrix-isolation literature.

Very many review articles relevant to matrix isolation have been written, both long and short. The bibliography mentioned above lists no fewer than 101 reviews appearing between 1958 and 1985, and more have appeared since. In view of the fact that this bibliography may not be as widely accessible as one should wish, a selection of some of the more general reviews follows.

Pimentel himself wrote some of the early reviews of various aspects of matrix isolation.[29,30] A guide to the literature on IR spectroscopy at sub-ambient temperatures was published in 1969 with 612 references,[31] but this covers quite a wide area, and more focused reviews on matrix vibrational spectroscopy appeared at about the same time[32] and later,[33-37] including extensive tabulations of data.[38] Three general reviews on chemical species isolated in matrices were published in *Molecular Spectroscopy* between 1973 and 1979,[39-41] but the series was not continued thereafter. Numerous specific areas of matrix chemistry have been reviewed, including studies of atoms,[42-49] small molecules,[50] ions,[51-55] metal carbonyls and organometallics,[56-60] as well as organic species.[20,21] It is hoped that this selective list of books and reviews will give the reader an adequate basis for further searches.

2. Matrix-isolation experiments—some general considerations

2.1 Why carry out matrix experiments in the first place?

The technique of matrix isolation was originally invented as a means of trapping reactive species and studying them spectroscopically with more or less conventional spectrometers (see Fig. 1.1). Although this remains its principal application, experience soon showed that matrix isolation has some advantages even for the study of stable molecules, and a surprising range of applications can now be found, for example in environmental analysis or in mimicking the conditions in interstellar dust clouds. For chemists, the most common reasons for wanting to do matrix experiments are to:

- observe directly and characterize reaction intermediates
- generate and study novel reactive species
- determine the structures of reactive species
- characterize molecular complexes and study weak interactions between species
- freeze out and study particular molecular conformations.

Matrix isolation provides chemists with a means of testing proposed reaction mechanisms or of comparing the computed structures of reactive molecules with those determined by experiment. It can also yield spectra of reaction intermediates, weak complexes, or particular conformations of stable molecules for correlation with the results from other experimental techniques, such as flash photolysis.

2.1.1 Matrix synthesis

Over the years there have been a number of attempts to develop matrix isolation as a synthetic tool. This is an attractive idea in cases where a reaction follows different pathways in low-temperature matrices and normal, ambient conditions. Matrix synthesis can, in principle, yield greater selectivity or even completely different products. There is no great difficulty in scaling up matrix equipment to cope with hundreds of milligrams or even gram quantities of guest materials, and ways exist for recovering any stable products of such scaled up matrix reactions. Nevertheless, none of the attempts to promote matrix synthesis seems to have been a great success.

2.2 Experimental constraints
2.2.1 Matrix deposition

Matrices are formed by deposition from the gas phase on to a cold window. Any species which is to be trapped in a matrix must therefore be volatile, at least to a small extent. There are two ways of mixing the guest species with the host gas.

(a) If the guest has an easily measurable vapour pressure, it can be mixed with the host gas on a vacuum line by standard manometric techniques. This produces a gas mixture of guest and host, usually of known proportions. The ratio of host to guest is known as the *matrix ratio*. The gas mixture is then allowed into the vacuum chamber of the cold cell at a controlled rate and is deposited as a solid matrix.

(b) If the guest has low volatility, it is usual to evaporate it from a side-arm attached to the vacuum chamber of the cold cell, while the host gas is deposited simultaneously. It is difficult to measure the matrix ratio in these conditions. The host gas can be passed directly over the solid or liquid guest material in the side-arm to assist volatilization, and the guest can be heated. Some guest materials such as alkali metals can be heated very strongly in a small furnace, but others will decompose thermally at quite low temperatures.

A major constraint in the use of matrix isolation, therefore, is that the intended guest material must be volatile at a temperature below its decomposition point.

There is in principle another approach to forming matrices: first dissolving the guest material in a liquefied gas, such as argon or nitrogen, and then freezing the resulting solution. This strategy does not work well, however. Liquified gases at their normal boiling points are very poor solvents for most materials, so only extremely dilute solutions can be prepared in this way. Moreover, in freezing such solutions, there is a risk of the guest molecules aggregating or even crystallizing out; so proper isolation of the molecules would be unlikely. In the future, this approach to matrix formation may become viable, with developments in the techniques for handling supercritical fluids, for example. Until then, really involatile materials such as polypeptides or polysaccharides are off limits for the matrix chemist.

2.2.2 Spectroscopic and analytical methods

A very wide range of spectroscopic techniques can be used with low-temperature matrices and, for the most part, fairly routine spectrometers will suffice. Nevertheless, for those coming to matrix isolation for the first time from a more general chemical research background, it is something of a shock to realize that NMR spectroscopy is, for all practical purposes, unavailable. It is true that a few research groups have developed special forms of NMR for matrices, but custom-built spectrometers are needed and the solid-state spectra obtained are of low resolution, lacking the coupling information which makes conventional NMR in liquids such a powerful structural tool.

For the most part, matrix chemists have to make do with IR and UV–visible spectroscopy, occasionally supplemented with less common techniques such as Raman spectroscopy, laser-induced emission and ESR (for species with unpaired electrons), or even more esoteric techniques such Mössbauer spectroscopy or magnetic circular dichroism, if appropriate. On the positive side, it is often possible to characterize a matrix isolated species by more than one form of spectroscopy.

i. Characterizing and identifying reactive species in matrices

When one has prepared a stable compound, preferably as a nice crystalline solid in a sample tube, it is an easy matter to collect all the data needed for its characterization and identification. It is only necessary to extract a few crystals of the sample and record a melting point or an IR spectrum, or send them off to the NMR, mass-spectrometry or microanalytical services. It would be a very poorly organized chemist indeed who, at the end, had any doubts that the spectroscopic and analytical data obtained all related to the same compound. What happens when a reactive species exists only in a low-temperature matrix, however? In these circumstances, we do well to obtain even two different types of spectrum for the species, and microanalysis is out of the question.

In the most favourable circumstances, it may be possible to obtain two different types of spectrum (e.g. IR and UV–visible absorption) with the same matrix sample. More often, because of the differing sensitivities of different types of spectroscopy and the incompatible requirements of different types of spectrometer for sample size and shape (e.g. ESR vs. IR), it is possible to obtain two or more types of spectrum only in separate matrix experiments. In any case, there is a problem in ensuring that two or more spectra do actually belong to one and the same species.

Another problem arises where, as is often the case, several reactive species may have been generated in a matrix, rather than just one. How is it possible to be sure that a set of IR bands, for example, all belong to a single molecular species? There is a basic rule of thumb in matrix isolation: bands which grow together at the same rate and diminish together at the same rate belong to the same species. Ideally, therefore, one generates a species of interest by as many different routes as possible. One also tries to find several ways of destroying it in the matrix, for example by warming to allow diffusion and bimolecular reaction or by photolysis. With careful work, one can become reasonably confident that a set of bands really does belong to a single species. The mis-assignment of a perturbed band of CO_2 to cyclobutadiene, which was discussed in Section 1.3, provides a good example of the pitfalls that can be encountered and how generating a species by more than one route can avoid them. This is by no means a unique example, however: the problem is quite general.

With the limitations in spectroscopic analysis that we have discussed, it is clear that the interpretation of spectroscopic data for reactive species in matrices requires care. Greater reliance has to be placed on IR spectroscopy than would be usual with stable compounds. The information to be derived from IR spectra can often be augmented by isotopic substitution. The resulting isotope shifts can be of decisive help in identifying reactive species and assigning bands to particular vibrations. It is often easy to replace hydrogen with deuterium in a precursor molecule, either generally or at specific positions, but incorporation of other isotopes, such as ^{13}C, ^{15}N or ^{18}O, may require long and expensive syntheses. It is not always feasible to undertake such isotopic substitution within a reasonable time or at reasonable cost. It has to be admitted, therefore, that reactive species in matrices will generally not be as fully characterized as would be expected for a stable compound.

In order to ensure that a matrix isolated species is identified as rigorously as possible, the following advice should be heeded.

(a) A novel reactive species should be generated by as many independent methods as possible and, in the resulting matrix spectra, any features which do not maintain constant relative intensities with respect to the others should not be assigned to that species.

(b) The relative intensities of assigned bands should also be checked in processes in which the reactive species is destroyed; and if more than one way of accomplishing this can be found, so much the better.

(c) Matrix experiments should be carried out with more than one host gas, to eliminate any specific matrix effects (see below).

(d) In experiments which rely on vibrational spectra, isotopic variants should be studied if they are accessible.

(e) As many different types of spectroscopy as possible should be employed.

ii. Recovering and analysing products from matrices

When a matrix is warmed up at the end of an experiment, it evaporates and is lost. It is sometimes possible to retrieve a guest species by pumping the gases from the evaporating matrix through a trap cooled with liquid nitrogen. The most reactive species will of course dimerize, or react in other ways, when the matrix softens on warming, but the resulting products may well be of interest and therefore worth trapping out. There is also sometimes an involatile solid residue left on the cold window after the matrix has evaporated. As often as not, this is an uninteresting polymer, but occasionally it is worth investigating. The quantities of guest material involved in a matrix experiment are usually very small (< 1 mg), but retrieval of products can be expected to provide enough sample at least for a mass spectrum.

2.2.3 The generation of reactive species

Broadly speaking, there are four ways in which reactive species may be generated and trapped in low-temperature matrices:

(a) deposition of a matrix containing a stable precursor, followed by irradiation of the matrix, for example with UV light, X-rays, or an electron beam;

(b) external generation of the reactive species in the gas phase, for example by photolysis, pyrolysis, or microwave excitation, followed by deposition in a stream of the host gas;

(c) co-condensation of two streams of material on to the cold window, for example (i) a beam of lithium atoms from a furnace and (ii) a mixture of an alkyl halide and the host gas, followed by reaction at the matrix surface during deposition;

(d) allowing a reactive species, already generated by one of the methods listed above, to undergo a thermal reaction within the matrix, either with a reactive host, such as CO, or with another guest species.

Examples of these four approaches are given in Scheme 1.3. Thermal reactions within a matrix rely on being able to warm the matrix to a softening point which allows the guest species to diffuse to a certain extent. This is called *annealing* the matrix and is usually accomplished by means of a resistance

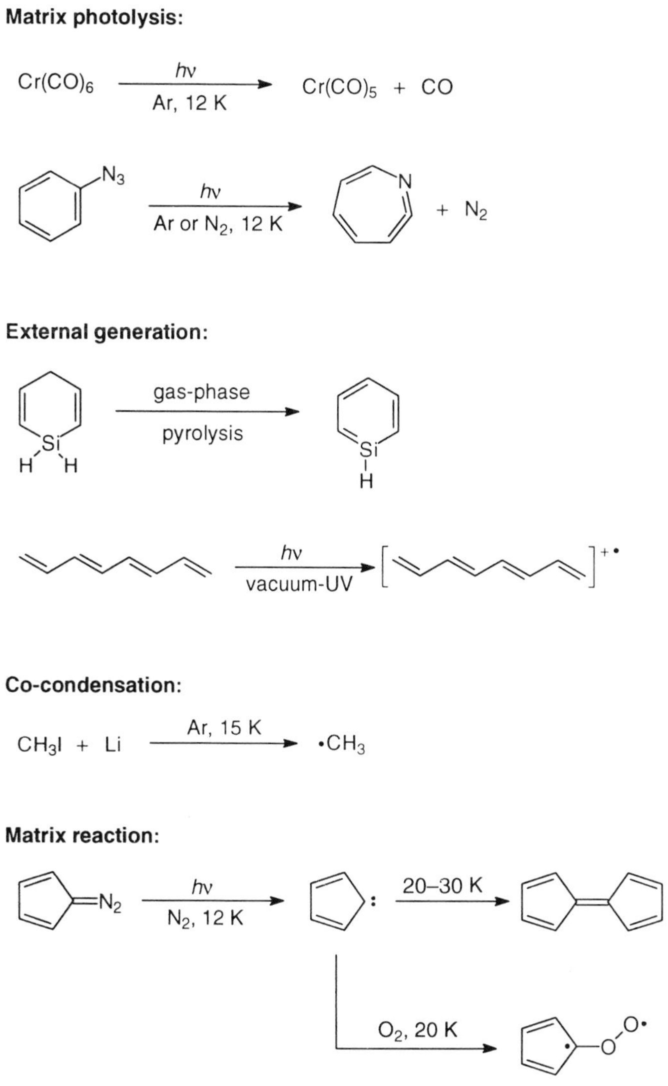

Matrix photolysis:

$$Cr(CO)_6 \xrightarrow[\text{Ar, 12 K}]{h\nu} Cr(CO)_5 + CO$$

$$\xrightarrow[\text{Ar or N}_2\text{, 12 K}]{h\nu} + N_2$$

External generation:

$$\xrightarrow[\text{pyrolysis}]{\text{gas-phase}}$$

$$\xrightarrow[\text{vacuum-UV}]{h\nu} \left[\right]^{+\bullet}$$

Co-condensation:

$$CH_3I + Li \xrightarrow{\text{Ar, 15 K}} \bullet CH_3$$

Matrix reaction:

$$=N_2 \xrightarrow[\text{N}_2\text{, 12 K}]{h\nu} : \xrightarrow{\text{20–30 K}}$$

$$\xrightarrow{O_2\text{, 20 K}}$$

Scheme 1.3

heater in thermal contact with the cold window holder. Diffusion of smallish molecules can usually be induced before wholesale evaporation of the matrix host begins. Argon and nitrogen matrices, for example, can be safely warmed to about 35 K for a few minutes, without appreciable loss.

i. Ions in matrices

The most common matrix hosts, solidified noble gases and nitrogen, are non-polar. Despite this, it is perfectly possible to generate ions in matrices. For example, radical cations can be generated from hydrocarbons by vacuum-UV photolysis during deposition.[52] The ejected electron is trapped by a suitable electron acceptor such as CCl_4. Radical anions can be generated by photo-ionization of Na atoms in the presence of an electron-acceptor molecule.[51] It is less easy to generate the more common types of (non-radical) carbocations and carbanions in matrices—one cannot carry out a matrix S_N1 reaction, for example—and so far there have been only a few matrix-isolation studies of these species.

2.2.4 Matrix effects

In any matrix-isolation study, it is important to give some consideration to the effects which the matrix host will have on the guest species. Fortunately, host–guest interactions in solidified noble gases tend to be weak; so perturbations of molecular structures or shifts in the frequencies of IR bands (from gas-phase values) can be expected to be small in most cases. The matrix host, however, does exert a profound influence on the diffusion and rotational motion of the guest species, and even the most inert host will interact electronically with trapped species if they are sufficiently reactive, for example metal atoms.

i. Physical effects

The most obvious effect of the matrix host is to prevent diffusion and rotation of the guest. The matrix host lattice acts not only as a cage but also as a clamp. Figure 1.9 shows gas-phase and matrix IR absorptions at about 1150 cm^{-1} belonging to the ν_1 vibration of SO_2 (the symmetric stretch). The gas-phase absorption has P and R branches with clearly resolved rotational fine structure, extending over more than 100 cm^{-1}. The matrix absorption, in contrast, is a fairly narrow band at 1151 cm^{-1} with a small sub-band at 1146 cm^{-1}; all rotational fine structure has been eliminated. In fact, the IR bands of matrix isolated molecules are often very narrow (< 1 cm^{-1}). This reduces the overlap of bands, especially when several different species may be present, and is a great help in analysing matrix IR spectra. It also increases the signal-to-noise ratio of the IR absorptions. For these reasons, matrix IR spectroscopy has been proposed for the routine analysis of gas and other volatile mixtures.

The small sub-band in Fig. 1.9(b) is due to a *matrix splitting* or *site effect*, and anyone carrying out matrix IR experiments will become familiar with this

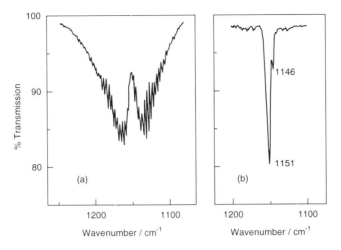

Fig. 1.9 IR spectra showing absorptions due to the symmetric stretching vibration of SO_2 near 1150 cm^{-1}: (a) in the gas phase at room temperature, and (b) matrix isolated in N_2 at 12 K (ratio N_2:SO_2 = 100:1). The spectra were recorded on the same spectrometer with 1.5 cm^{-1} resolution.

phenomenon, which is very common. Owing to the naturally small bandwidths of matrix IR spectra, band splittings of various types become more apparent than in conventional liquid- or solid-phase IR spectra. The three main causes of band splitting in the IR spectrum are

• Fermi resonance

• variable host–guest interactions

• guest-guest interactions, that is, molecular aggregation.

Fermi resonance is an intrinsic property of the molecule, but is often especially apparent in matrix spectra, owing to the small bandwidths. Variable inter-actions between the host and guest arise when the guest can occupy more than one type of vacancy in the host lattice. This situation is depicted in Fig. 1.1, where one of the trapped guests occupies a vacancy left by three host atoms, while the others are in larger matrix cages. Finally, aggregation of the guest molecules can result in various forms of guest–guest interaction, ranging from loose van der Waals association, through dipole–dipole alignment, to strong hydrogen-bonding or other complexation. Whenever a guest species can exist in the matrix in a variety of different sites, band splitting is likely to arise.

It is not quite true to say that the matrix host will always stop guest molecules rotating. For small molecules, including SO_2, there is evidence that hindered rotation can still occur within the matrix cage. The trapped molecules rotate more freely as the temperature is raised, and this sometimes shows up as a *reversible* variation with temperature of the relative intensities of components

19

of a split band. This phenomenon is discussed in Hallam's book,[25] but is not expected to arise with the majority of matrix-isolated species.

It is not always necessary to be concerned about the causes of observed band splittings, but in the most exacting studies it is always preferable to identify splittings and their origins, and thus avoid assigning split sub-bands to separate fundamental vibrations. Host–guest interactions can usually be identified by obtaining spectra of a species of interest in more than one matrix host. The cavity sizes in Ne, Ar, Kr, and Xe matrices, for example, vary considerably, and band splittings due to host–guest interactions will also vary as a consequence. Band splitting due to molecular aggregation can usually be identified by varying the matrix (host:guest) ratio. At sufficiently high dilution, the trapped molecules will be well isolated, and guest–guest interactions minimized.

ii. Chemical effects

Apart from the physical restraint which the matrix host exerts on guest species and the consequent effects on the matrix spectra, there is also the possibility of chemical interaction between host and guest. In some matrix experiments, a reactive host such as CO is chosen deliberately, but in the majority of cases a chemically inert host is required. There are some borderline choices. For example, nitrogen is cheap and experience shows that it gives nicely trans-parent matrices. It is popular as a matrix host for these reasons. It is chemically inert towards organic radicals and many other species, but coordinates strongly with metal atoms or metal complexes with vacant coordination sites. The safest choices when unreactive matrix hosts are required are therefore the noble gases, but even these are not completely inert.

A good example of a strong host–guest interaction involving the noble gases is provided by the case of matrix isolated $Cr(CO)_5$. This reactive species is conveniently generated in matrices by photolysis of $Cr(CO)_6$ (Scheme 1.3). The position of the maximum in the electronic absorption of $Cr(CO)_5$ has been found to vary greatly with different matrix hosts (Table 1.1).[61] Clearly the photoproduct in each case is not the naked $Cr(CO)_5$ molecule, but a complex between $Cr(CO)_5$ and the host species: $X \cdots Cr(CO)_5$ (X = Ne, Ar, Kr, Xe, CH_4, CF_4, SF_6). Neon shows the weakest interaction, which may be virtually negligible, but all the other host species coordinate significantly with the reactive guest.

When there is a choice of matrix host, neon will always be expected to show the weakest host–guest interactions. For this reason, it is regarded as the best choice whenever its use is possible. The deposition of neon matrices requires temperatures of 6 K or below, however, and the most popular commercial cryostats operate down to only about 10 K. Therefore neon matrices have not yet become routine for the majority of matrix laboratories, who do not have access to lower-temperature cryostats. At present, cryostats operating at 6 K or below are becoming more readily available, and much early work, originally

Table 1.1. Variations in the position of the visible electronic absorption of $Cr(CO)_5$ with matrix host

Matrix host	Ne	SF_6	CF_4	Ar	Kr	Xe	CH_4
λmax / nm[a]	624	560	547	533	518	492	489

[a] Data from ref. 61.

carried out with argon matrices, is being repeated in the less strongly interacting neon.

iii. Cage effects

Cage effects are well known to influence reaction pathways. The classic example is encountered when a pair of free radicals is generated in a viscous solvent such as decalin. In these conditions the two radicals have a high probability of reacting together, by dimerizing or disproportionating, before either has a chance to diffuse away from the other. With the lattice rigidity of low-temperature matrices, cage effects can be expected to be even more prevalent than in viscous fluids, and this is indeed the case. Fortunately, there are a number of factors which can come into play to reduce the likelihood of cage recombination in matrix experiments, otherwise it might be difficult to generate some types of reactive intermediate at all.

Matrix cage effects are most likely to arise in experiments in which a pair of potential reaction partners are generated in the same matrix cage by photolysis. In many cases, there is little or no chance of a reverse reaction for thermodynamic reasons. The photolytic generation of cyclobutadiene and CO_2, for example, leads to cage pairs of these two molecules (Scheme 1.2), but the photolysis is irreversible; so, although the IR spectra of the two product species may be perturbed by mutual interaction within the cage, there is no chance of a back reaction. This is usually the case when small very stable molecules, such as N_2 or CO_2, are eliminated in matrix photolyses.

When any photolysis takes place in a matrix, there is a great deal of excess vibrational energy to be dissipated after the reaction has occurred. This can allow cage pairs to separate and thus prevent any possibility of back reaction. The photolysis of $Cr(CO)_6$, for example, generates $Cr(CO)_5$ and a molecule of CO, and the reverse thermal reaction can occur readily in the matrix. If the CO molecule escapes the matrix cage in which it was generated, however, the reverse reaction is prevented, and $Cr(CO)_5$ persists as a trapped species. There are two ways in which such an escape from the matrix might occur. The eliminated small molecule may have sufficient energy to blast out of the matrix cage, without softening the host lattice significantly, like a bullet being fired into a mattress. Alternatively, the excess vibrational energy arising in the reaction might soften a region of the matrix around the reaction centre, and thus allow momentary diffusion of the products, until the host lattice is cooled and becomes rigid again. This latter idea is often referred to as the *local soup theory*. Whatever the exact mechanism, the immediate separation of cage pairs following their formation will inhibit a reverse reaction and allow the reactive species to persist in the matrix.

The situation with metal carbonyls may be more complicated, however. There is evidence to suggest that, even when an ejected molecule of CO remains in the same cage as the metal carbonyl fragment, recombination will not occur if the metal carbonyl rotates so as to have its vacant coordination site facing away from the CO molecule. The metal carbonyl fragment, in effect, turns its back on the CO molecule, and this is enough to prevent recombination.

Another way in which thermally favourable cage recombinations are prevented is when the potential reaction partners are generated with a third, unreactive species separating them. A good example is the photolysis of dipropanoyl peroxide in Ar matrices (Scheme 1.4). This reaction produces a pair of ethyl radicals in the matrix cage, but they are separated by the two molecules of CO_2 which are also formed, and so the radicals persist and can be observed by IR spectroscopy. On warming the matrix to 30 K, the CO_2 molecules diffuse away and the ethyl radicals dimerize to give butane, or disproportionate to molecules of ethane and ethylene.

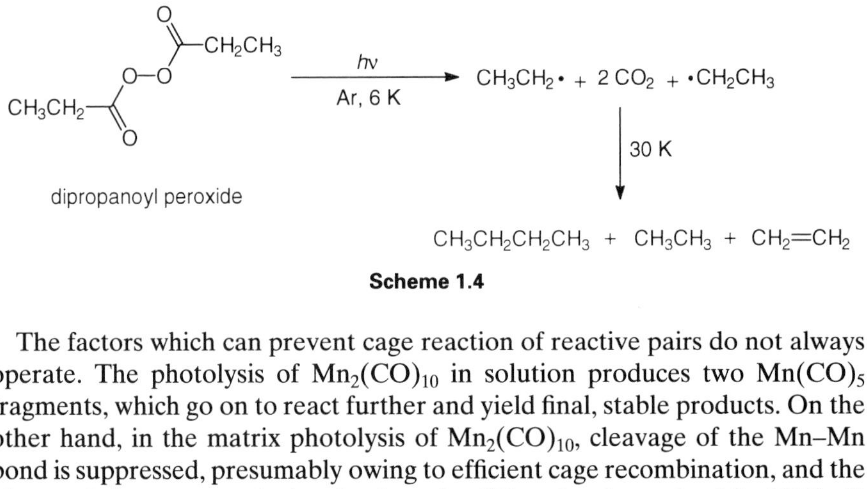

$$CH_3CH_2\cdot + 2\,CO_2 + \cdot CH_2CH_3$$

$$\downarrow \;30\,K$$

$$CH_3CH_2CH_2CH_3 + CH_3CH_3 + CH_2{=}CH_2$$

Scheme 1.4

The factors which can prevent cage reaction of reactive pairs do not always operate. The photolysis of $Mn_2(CO)_{10}$ in solution produces two $Mn(CO)_5$ fragments, which go on to react further and yield final, stable products. On the other hand, in the matrix photolysis of $Mn_2(CO)_{10}$, cleavage of the Mn–Mn bond is suppressed, presumably owing to efficient cage recombination, and the observed photoprocess is the ejection of a CO molecule to give $Mn_2(CO)_9$. It seems that CO is small enough to escape from the matrix cage or migrate to an unreactive side of the $Mn_2(CO)_9$ molecule, but the $Mn(CO)_5$ fragments simply remain trapped together in the cage.

2.3 Matrix isolation in relation to other techniques

Amongst research groups who have their own matrix-isolation equipment, there are some who do nothing but matrix experiments while others use matrix-isolation as just one of many techniques. In any investigation of reaction mechanisms or reactive intermediates, there will usually be a choice of experimental approach. Whether matrix studies represent a worthwhile option depends on the chemistry to be studied. Research on carbocations, for example, has been pursued much more effectively by NMR spectroscopy of superacid solutions than by matrix IR spectroscopy. In many cases, a combination of techniques will be desirable. The discussions in this chapter are intended to give a broad guide to the circumstances in which matrix experiments are likely to yield valuable results.

A frequently asked question in photochemical research is whether matrix isolation or flash photolysis offers the best approach to understanding a reaction pathway. Table 1.2 presents a comparison of these two techniques, from which it can be seen that, in several respects, they are complementary. Flash-photolysis experiments have the advantage of giving the lifetimes of reactive species, but, until recently, the observation of transient species in flash-photolysis studies was virtually limited to UV–visible absorption, which does not provide very much structural information on which to base an identification. In contrast, matrix-isolation studies of the same photoreaction could link the UV–visible absorption of a reactive intermediate to a set of IR bands, thus giving a more assured basis for identification. The development of faster IR detectors has meant that transient species can now be observed in the

---

Table 1.2. Comparison of flash photolysis and matrix isolation

Flash photolysis	Matrix isolation
Reactions in gas, liquid or solid phases.	Solid-phase reactions only, but precursor has to be volatile.
Reactive species are observed in the normal conditions of a reaction medium.	Reactive species are stabilized in the abnormal conditions of a matrix cage.
Reactive species are transient.	Reactive species are long lived—often indefinitely.
Spectra are usually recorded at one wavelength per flash — multiple flashes needed.	Whole spectrum is recorded normally, and, in favourable cases, different types of spectra can be recorded for the same matrix.
Detection of transient species is by UV–visible or IR absorption, but only part of the IR region is accessible at present.	Many types of spectroscopy are applicable: including far-, mid- and near-IR absorption; Raman, UV-visible absorption, ESR, Mössbauer, and various type of emission spectroscopy.
Reactive species are generated by photolysis.	Reactive species can be generated by a range of methods: including photolysis, vacuum pyrolysis, microwave discharge, electron bombardment and chemical reaction.

IR in flash studies, but still over only a limited wavenumber range. The useful symbiosis of matrix isolation and flash photolysis is likely to continue a little while longer, therefore.

Finally, a word of caution is in order. Anyone gaining access to a matrix-isolation laboratory will quickly discover that matrix experiments can be very convenient to carry out—far easier than messy preparative work. There is always a temptation to carry out matrix experiments on a novel 'reactive' species, before finding out just how reactive it really is. Hence there is a danger of generating a new species in a matrix when it is stable enough to be prepared in a glove box at room temperature, or even in the open air. The author does not know of any actual examples of this happening, but confesses that, on some occasions in his own laboratory, the room-temperature control reactions have been carried out long after the matrix experiments.

There is, however, one spectacular case, a chemical triumph rather than a *faux pas*, in which a molecule generated in matrices because of its anticipated instability, turned out, in the end, to be surprisingly stable. This was the synthesis of tetra-*t*-butyltetrahedrane. The tetrahedrane skeleton was calculated to have about 550 kJ mol^{-1} of strain energy, and theoretical predictions for the chances of synthesizing a tetrahedrane derivative of any sort were uniformly gloomy. Matrix isolation probably offered the best hope. Surprisingly, Günther Maier and co-workers discovered that the tetra-*t*-butyl derivative could be generated in Ar matrices by photolysis of tetra-*t*-butylcyclo-pentadienone, the reaction proceeding via a criss-cross cycloadduct followed by expulsion of CO (Scheme 1.5).[62,63]

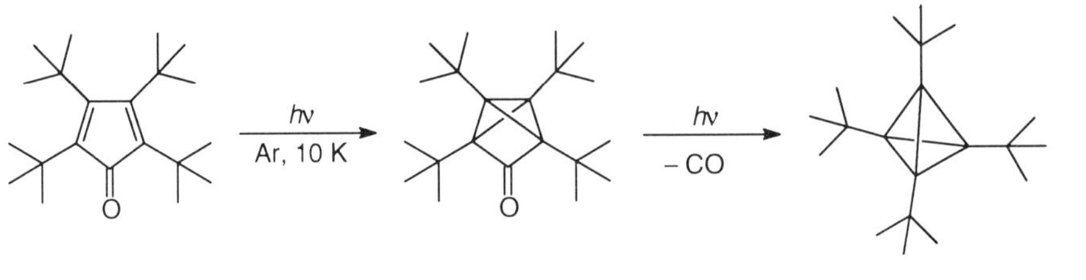

tetra-*t*-butyltetrahedrane

Scheme 1.5

As would be expected, matrix IR spectra were not enough to identify the tetrahedrane, but fortunately the same photochemistry could be carried out in a frozen solvent at 77 K, or in liquid solutions between $-130°C$ and room temperature, thus permitting analysis by ^{1}H and ^{13}C NMR. Astonishingly, it was eventually discovered that the tetrahedrane could be isolated from these solutions by chromatography, and was stable up to about $+130°C$. Its structure was ultimately confirmed by X-ray crystallography.[64] Few matrix-isolation studies will end thus.

References

1. Lewis, G. N.; Lipkin, D. *J. Am. Chem. Soc.* **1942**, *64*, 2801–2808.
2. Michaelis, L.; Schubert, M. P.; Granick, S. *J. Am. Chem. Soc.* **1939**, *61*, 1981–1992.
3. Lewis, G. N.; Lipkin, D.; Magel, T. T. *J. Am. Chem. Soc.* **1944**, *66*, 1579–1583.
4. See, for example, Porter, G.; Strachan, E. *Spectrochim. Acta* **1958**, *12*, 299–304.
5. Reiser, A.; Fraser, V. *Nature* **1965**, 682–683.
6. Reiser, A.; Bowes, G.; Horne, R. J. *Trans. Faraday Soc.* **1966**, *62*, 3162–3169.
7. Reiser, A.; Wagner, H. M.; Marley, R.; Bowes, G. *Trans. Faraday Soc.* **1967**, *63*, 2403–2410.
8. Reiser, A.; Marley, R. *Trans. Faraday Soc.* **1968**, *64*, 1806–1815.
9. Leyva, E.; Platz, M. S.; Persy, G.; Wirz, J. *J. Am. Chem. Soc.* **1986**, *108*, 3783–3790.
10. Platz, M. S. *Acc. Chem. Res.* **1995**, *28*, 487–492.
11. Dunkin, I. R.; Lynch, M. A.; McAlpine, F.; Sweeney, D. *J. Photochem. Photobiol. A: Chem.* **1997**, *102*, 207–212.
12. Bass, A. M.; Broida, H. P., eds. *Formation and Trapping of Free Radicals*; Academic Press: New York, **1960**.
13. Meyer, B. *Low Temperature Spectroscopy*; Elsevier: New York, **1971**.
14. Whittle, E.; Dows, D. A.; Pimentel, G. C. *J. Chem. Phys.* **1954**, *22*, 1943.
15. See, for example, Jen, C. K.; Bowers, V. A.; Cochrane, E. L.; Foner, S. N. *Phys. Rev.* **1962**, *126*, 1749–1757.
16. Lin, C. Y.; Krantz, A. *J. Chem. Soc., Chem. Commun.* **1972**, 1111–1112, 1316.
17. Chapman, O. L.; McIntosh, C. L.; Pacansky, J. *J. Am. Chem. Soc.* **1973**, *95*, 244–246.
18. Pong, R. G. S.; Shirk, J. S. *J. Am. Chem. Soc.* **1973**, *95*, 248–249.
19. Masamune, S.; Sonto-Bachiller, F. A.; Machiguchi, T.; Bertie, J. E. *J. Am. Chem. Soc.* **1978**, *100*, 4889–4891.
20. Dunkin, I. R. *Chem. Soc. Rev.* **1980**, *9*, 1–23.
21. Dunkin, I. R. In *Chemistry and Physics of Matrix Isolated Species*; Andrews, L.; Moskovits, M. eds; North-Holland: Amsterdam, **1989**, Ch. 8, pp. 203–237.
22. Andrews, W. L. S.; Ault, B. S.; Berry, M. J.; Moore, C. B. *J. Phys. Chem.* **1991**, *95*, 2607–2615.
23. Cradock, S.; Hinchcliffe, A. J. Matrix Isolation: a Technique for the Study of Reactive Inorganic Species; Cambridge University Press, **1975**.
24. Moskovits, M.; Ozin, G. A., eds. *Cryochemistry*; Wiley: New York, **1976**.
25. Hallam, H. E., ed. *Vibrational Spectroscopy of Trapped Species*; Wiley: London, **1973**.
26. Barnes, A. J.; Orville-Thomas, W. J.; Müller, A.; Gaufrès, R., eds. *Matrix Isolation Spectroscopy*; Reidel: Dordrecht, **1981**.
27. Andrews, L.; Moskovits, M., eds. *Chemistry and Physics of Matrix Isolated Species*; North-Holland: Amsterdam, **1989**.
28. Ball, D. W.; Kafafi, Z. H.; Fredin, L.; Hauge, R. H.; Margrave, J. L. *A Bibliography of Matrix Isolation Spectroscopy: 1954–1985*; Rice University Press: Houston, **1988**.
29. Pimentel, G. C. *Pure Appl. Chem.* **1962**, *4*, 61–70.
30. Pimentel, G. C.; Charles, S. W. *Pure Appl. Chem.* **1963**, *7*, 111–123.
31. Hermann, T. S.; Harvey, S. R. *Appl. Spectrosc.* **1969**, *23*, 435–467.
32. Barnes, A. J.; Hallam, H. E. *Quart. Rev.* **1969**, *23*, 392–409.
33. Andrews, L. *Ann. Rev. Phys. Chem.* **1971**, *22*, 109–132.

34. Nibler, J. W. *Adv. Raman Spectrosc.* **1972**, *1*, 70–75.
35. Knözinger, E.; Jacox, M. E. *Ber. Bunsenges. Phys. Chem.* **1978**, *82*, 57–60.
36. Knözinger, E.; Wittenbeck, R. *Infrared Phys.* **1984**, *24*, 135–142.
37. Guenthard, H. H. *J. Mol. Struct.* **1984**, *113*, 141–145.
38. Jacox, M. E. *J. Phys. Chem. Ref. Data.* **1984**, *13*, 945-1068.
39. Downs, A. J.; Peake, S. C. *Mol. Spectrosc.* **1973**, *1*, 523–607.
40. Chadwick, B. M. *Mol. Spectrosc.* **1975**, *3*, 281–382.
41. Chadwick, B. M. *Mol. Spectrosc.* **1979**, *6*, 72–135.
42. Ozin, G. A.; Vander Voet, A. *Acc. Chem. Res.* **1973**, *6*, 313-318.
43. Kündig, E. P.; Moskovits, M.; Ozin, G. A. *Angew. Chem., Int. Edn. Engl.* **1975**, *14*, 292–303.
44. Pimentel, G. C. *Angew. Chem., Int. Edn. Engl.* **1975**, *14*, 199–206.
45. Turner, J. J. *Angew. Chem., Int. Edn. Engl.* **1975**, *14*, 304–308.
46. Ozin, G. A.; Vander Voet, A. *Progr. Inorg. Chem.* **1975**, *19*, 105–172.
47. Ozin, G. A. *Acc. Chem. Res.* **1977**, *10*, 21–26.
48. Perutz, R. N. *Chem. Rev.* **1985**, *85*, 77–96.
49. Ozin, G. A.; McCaffrey, J. G.; Parnis, J. M. *Angew. Chem., Int. Edn. Engl.* **1986**, *25*, 1072–1085.
50. Perutz, R. N. *Chem. Rev.* **1985**, *85*, 97–127.
51. Kasai, P. H. *Acc. Chem. Res.* **1971**, *4*, 329–336.
52. Andrews, L. *Ann. Rev. Phys. Chem.* **1979**, *30*, 79–101.
53. Miller, T. A.; Bondybey, V. E. *Appl. Spectrosc.* **1982**, *18*, 105–169.
54. Shida, T.; Haselbach, E.; Bally, T. *Acc. Chem. Res.* **1984**, *17*, 180–186.
55. Haselbach, E.; Bally, T. *Pure Appl. Chem.* **1984**, *56*, 1203–1213.
56. Turner, J. J.; Burdett, J. K.; Perutz, R. N.; Poliakoff, M. *Pure Appl. Chem.* **1977**, *49*, 271–285.
57. Poliakoff, M. *Chem. Soc. Rev.* **1978**, *7*, 527–540.
58. Burdett, J. K. *Coord. Chem. Rev.* **1978**, *27*, 1–58.
59. Hitam, R. B.; Mahmoud, K. A.; Rest, A. J. *Coord. Chem. Rev.* **1984**, *55*, 1–29.
60. Almond, M. J. *Chem. Soc. Rev.* **1994**, *23*, 309–317.
61. Perutz, R. N.; Turner, J. J. *J. Am. Chem. Soc.* **1975**, *97*, 4791–4800.
62. Maier, G.; Pfriem, S.; Schäfer, U.; Matusch, R. *Angew. Chem., Int. Edn. Engl.* **1978**, *17*, 520–521.
63. Maier, G.; Pfriem, S.; Schäfer, U.; Malsch, K.-D.; Matusch, R. *Chem. Ber.* **1981**, *114*, 3965–3987.
64. Irngartinger, H.; Goldmann, A.; Jahn, R.; Nixdorf, M.; Rodewald, H.; Maier, G., Malsch, K.-D.; Emrich, R. *Angew. Chem., Int. Edn. Engl.* **1984**, *23*, 993–994.

2

Equipment: the matrix-isolation cold cell

1. Introduction

For matrix-isolation studies, there is a minimum of necessary equipment, and the complexity and cost of this may initially seem daunting. The essentials include

- a refrigeration system (cryostat)
- a sample holder
- a vacuum chamber (shroud) to enclose the sample
- a means of measuring and controlling the sample temperature
- a vacuum-pumping system
- a gas-handling system
- methods of generating the species of interest (e.g. UV irradiation or pyrolysis)
- methods of analysing the matrices (usually with one or more spectrometers).

This chapter describes the major equipment needed for a matrix-isolation unit capable of performing various types of experiments, and gives designs of basic systems and advice on construction and maintenance. Chapter 3 deals with ancillaries, such as preparative vacuum lines, photolysis sources and spectrometers. All the items in the list above are discussed, together with some related topics. Descriptions of additional items of minor or ancillary equipment will be found in other chapters at the points where they are most relevant. Although as much helpful detail as possible has been included, there are as many ways of putting together a system as there are laboratories. Readers should thus regard the advice given about equipment as a general guide to what is needed rather than a set of immutable instructions. It is hoped that the discussion which follows will stimulate those interested in setting up a matrix laboratory to find their own most appropriate solutions.

Descriptions of equipment for matrix isolation have appeared in a number of earlier books,[1-6] and these are recommended as additional sources of design ideas. The primary aim of this chapter is to enable research workers new to matrix isolation to set up a basic system from scratch and have it running as quickly as possible. Many designs of cold cells for specific types of matrix experiments, some of them very complex, are to be found in the research literature, and these should be sought out and studied if a more complicated system is needed. In any case, contact with research groups already experienced in matrix techniques should prove valuable, and opportunities to inspect and discuss working matrix-isolation systems should always be taken. Suppliers are invariably helpful both with direct advice and by establishing contacts with other users of their equipment.

2. Refrigerators

2.1 Bath and flow cryostats

The first work on matrices made use of refrigeration by liquid helium (b.p. 4.2 K) or liquid hydrogen (b.p. 20.3 K). The liquified refrigerants were stored in Dewar vessels and transferred to specially designed sample cells—bath cryostats—to maintain a fairly constant coolant level (see Chapter 1). Unless multiple storage Dewars were available, the maximum duration of matrix experiments with these liquified gases depended on how long the contents of the full storage Dewar (usually 50 or 100 litres) would last: in practice a few hours.

Although relatively inexpensive, liquid hydrogen is much too inflammable to be recommended as a practical coolant. Fires sometimes occurred when hydrogen, evaporating from a matrix-isolation cell, contacted the glowing source of an IR spectrometer. Reputedly on some occasions research workers heroically carried on recording important spectra with flames emanating from the cold cell. Liquid helium is much safer than liquid hydrogen but is expensive. It does have the advantage of a very low boiling point, however, and is still used in certain types of matrix experiment, in which the consumption of helium is not excessive.

Today there are several manufacturers of cryostats using liquid helium as the cryogenic fluid. These are either bath cryostats or flow cryostats, the latter operating with a continuous flow of the cryogen. Which type of cryostat should be chosen depends primarily on the duration of the experiments to be conducted. Flow cryostats are preferable for experiments lasting up to a few hours, while bath cryostats are better for longer experiments. Modern liquid-helium cryostats are very efficient and are advantageous when:

- a temperature below 10 K is required
- a very compact cryostat is necessary
- sample vibration must be avoided.

When a liquid-helium cryostat is used, at least one liquid-helium storage Dewar is needed together with transfer lines. It is also desirable to recover the helium in a large balloon and then compress it into cylinders for return to the supplier. The recovery of helium in this way is inconvenient and requires additional plant, particularly a suitable helium compressor.

2.2 Closed cycle helium refrigerators

The development of closed cycle helium refrigerators brought to an end the routine use of liquefied gases as coolants for most matrix experiments. These refrigerators consist of a compressor unit connected to a compact expander unit or head module by high pressure (feed) and low pressure (return) helium lines. The head module is small and light enough to be incorporated into matrix cells for use with a wide variety of instrumentation. The helium is compressed and then allowed to expand within the head module, most commonly in two stages. The expansion of the helium produces the cooling effect.

Figure 2.1 gives a simplified view of the internal parts of a typical two-stage head module and Fig. 2.2 shows the flow of helium during the intake and exhaust strokes. The flow of helium is controlled by a continuously rotating valve, which is driven by an electric motor. Below the valve is a slack cap and displacer, the latter consisting of two regenerators coupled together. As its name implies, the slack cap can move vertically with respect to the displacer. Both the slack cap and the displacer move up and down within the head module in response to changes in helium pressure. On the intake stroke, helium is admitted through passages in the slack cap and enters the regenerators. The regenerators, cooled during the previous cycle, cool the incoming gas. The gas under pressure lifts the slack cap and the displacer. As the displacer rises, gas above the slack cap is further compressed and passes into the surge volume. At the end of the intake stroke, the compressed gas in the surge volume acts as a buffer and stops the displacer striking the valve stem. On the exhaust stroke, high pressure gas at the two heat stations is free to expand. The expansion cools both the heat stations and the regenerators. As the pressure decreases, gas in the surge volume pushes the slack cap and displacer downwards, ready for the next cycle.

Because the gas remains under pressure, the lowest temperature attainable by two-stage closed cycle refrigerators at the second heat station—about 10 K—is higher than the boiling point of helium at atmospheric pressure. Nevertheless, the attainable temperatures are adequate for most matrix-isolation needs. In a typical two-stage module, the first heat station is maintained at about 80 K, and it is usual to fit it with a shield to protect the second stage from incoming radiation (see Section 3.4).

Closed cycle helium refrigerators are expensive initially, but once purchased are cheap to run. In a well maintained set-up, the helium might need to be topped up once a year, possibly even less frequently, while the power consumption of a typical unit is only 1.8 kW. Closed cycle refrigerators can run for thousands of hours with minimal maintenance. With such a refrigerator the

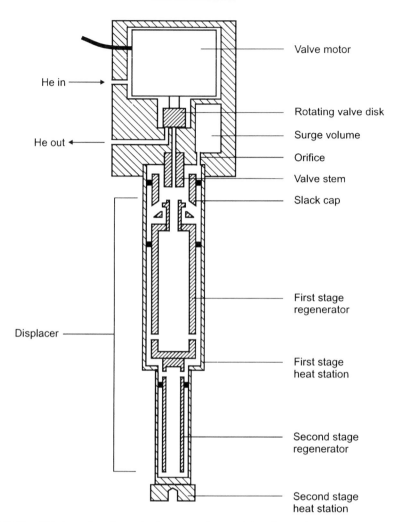

Fig. 2.1 Simplified section through the expander module of a typical two-stage closed cycle helium refrigerator.

sample holder of a matrix-isolation cell can be cooled from room temperature to 10 K in an hour or less. Compared with flow cryostats, closed cycle refrigerators are bulky and noisy. Also, owing to the reciprocating action of the displacer, a small amount of vibration can usually be felt at the head module during operation, though this is of no consequence in most matrix-isolation experiments. Closed cycle refrigerators require only an electrical supply and cooling water, no helium-storage Dewars or recovery plant. Even the cooling water can be dispensed with if an air-cooled compressor unit is specified, but these give out appreciable amounts of heat to the laboratory and

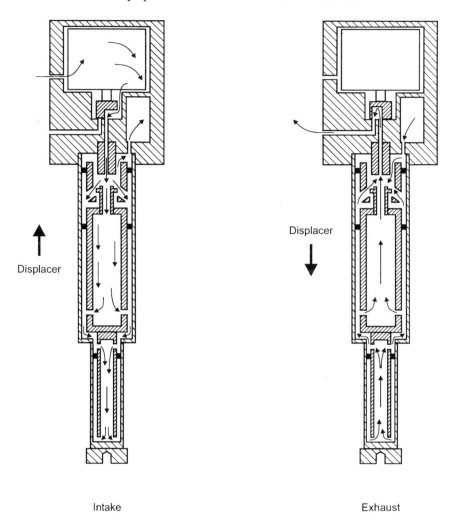

Displacer

Displacer

Displacer

Intake

Exhaust

Fig. 2.2 Flow of helium through the expander module on the intake and exhaust strokes.

should be avoided if possible. The convenience of use and low running costs of closed cycle refrigerators far outweigh any disadvantages in nearly all matrix applications. Only if temperatures below 10 K or other special conditions are required will the possibility of a liquid-helium cryostat be worth considering. Even then, three-stage closed cycle refrigerators now exist which can cool down to about 4 K. These are, however, very expensive and the head modules tend to be heavier and bulkier than those of two-stage refrigerators, though more compact designs are beginning to appear. A very recent development is a two-stage module capable of cooling to about 6 K.

When a matrix-isolation system is being designed, the lowest necessary temperature must be determined at the earliest stage, as must any special requirements, such as compactness or the need to carry out ESR experiments. For a general matrix-isolation system, a commercial two-stage closed cycle helium refrigerator is recommended. In this book, it is assumed that such a closed cycle refrigerator will be used, but most of the descriptions of methods and experiments will readily be adaptable to a bath or flow cryostat system. Suppliers of all types of cryostats are listed in the Appendix.

2.3 Ultra-low temperatures

Ultra-low temperatures are mostly the province of physicists. Cryostats capable of operating below 4 K are unnecessary for all but the most esoteric of matrix-isolation experiments.

With fairly standard rotary pumps, liquid helium can be pumped under reduced pressure to its λ-point at about 2 K, when it becomes a superfluid, and with further pumping to about 1.3 K. With much faster booster pumps temperatures below 1 K can be achieved. Commercially available cryostats using ^{3}He as the cryogenic fluid in the last stage can produce temperatures down to 0.3 K, while dilution refrigerators, utilizing mixtures of ^{3}He and ^{4}He, can get down to a few mK. The technique of adiabatic demagnetization, which does not make use of a cryogenic fluid, is capable of achieving temperatures in the μK range. Cryostats based on these principles will not be discussed further in this book, but interested readers can find more details elsewhere,[7] and some suppliers of more esoteric cryostats are listed in the Appendix.

3. Sample holders

Matrices are laid down on cold surfaces (substrates), which are chosen to suit the spectroscopic techniques to be employed. The sample holders are screwed into the lower heat station of the head module of the refrigerator. It is important to achieve good thermal contact between the surface on which the sample is to be deposited and the cold end of the refrigerator head module. Many materials that are good conductors of heat at room temperature become effective thermal insulators at very low temperatures. Copper is the most cost effective material for the metal parts of the sample holder, while indium gaskets are used to maximize thermal contact between parts. Screws holding the parts together need to be tightened to an extent which novices often find alarming, and need to be checked and retightened after the first few cooling cycles. A slightly loose sample holder will allow a significant temperature gradient to exist between the cold end of the refrigerator and the sample itself. This will result in poor quality matrices and excessive evaporation of the matrix host gas.

In the sections below, the specific requirements for particular forms of spectroscopy are discussed. Most matrix-isolation experiments use only one spectroscopic technique, but there are occasions when it is desirable to examine a matrix by more than one form of spectroscopy, most commonly IR and UV–visible. In these circumstances the choice of substrate will be especially critical.

3.1 IR and UV–visible absorption spectroscopy

3.1.1 CsBr and CsI cold windows for IR spectroscopy

For IR absorption spectroscopy, a good choice of substrate is a CsBr or CsI window of 25 mm diameter, in a copper or nickel-plated copper window holder screwed into the cold end of the refrigerator head module (Fig. 2.3(a)). The window is sandwiched between the two sections of the holder, which are screwed together. The crystal structures of CsBr and CsI are such that windows made of these materials have no easy cleavage planes, unlike KBr or NaCl windows, and are moreover rather soft and easily deformed. It is thus possible to tighten the screws of the window holder without risk of fracture. The best practice is to seat the window on a ring of indium wire, at least on the side of the holder which bears the mating thread, but this refinement is often omitted with little apparent detriment. The recommended procedure for fitting a cold window is given in Protocol 1.

Over a period, CsBr and CsI windows tend to develop thinned edges and a central bulge, rather like a lens. Also, because of their softness, it is not easy to

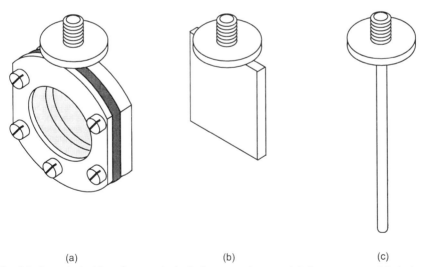

(a)   (b)   (c)

Fig. 2.3 Sample holders for matrix-isolation experiments. (a) A copper or nickel-plated copper holder with a salt window (e.g. CsBr or CaF$_2$) for IR or UV–visible transmission spectroscopy; (b) a copper or gold-plated copper plate for Raman or reflection spectroscopy; (c) a copper rod for ESR spectroscopy. Each holder has a mating thread to suit the cold tip of the refrigerator head module.

bring CsBr and CsI windows to a high degree of polish. Thus CsBr and CsI windows that have been in use for a long period appear optically imperfect. This does not usually impair their use for IR matrix studies, however.

Protocol 1.
Fitting a CsBr or CsI cold window in a matrix-isolation cell

The following procedure can be used for windows made of other materials, such as fused silica or CaF_2, with the important modification that, with these more brittle materials, the window holder must not be screwed up too tightly.

Caution!   Handle the windows carefully, only by the edges and preferably with thin rubber gloves. Place them down only on scrupulously clean surfaces.

Equipment
• A window holder to fit the head module of the refrigerator

Materials
• A polished CsBr or CsI window to fit the holder   **Soft, very hygroscopic**
• Indium wire
• Indium foil (a piece 1 cm square will do)

Indium wire and foil are usually supplied with refrigerator systems in quantities enough to last a few years. A small bottle of 'cryoconducting grease' is also often supplied, which is supposed to enhance thermal contact between the various components of the window assembly, but the author is unconvinced of its benefits.

1. Disassemble the window holder, keeping the screws and any washers together in a place where they will not roll out of sight.

2. Check that the window will fit into its recess in the holder. If it is slightly too big, use abrasive paper round the edges to reduce its size slightly. Try to avoid particles of the abrasive alighting on the polished surfaces of the window; remove any that fall there by tapping or blowing dry air, not by wiping.

3. Measure a length of indium wire to fit right around the groove at the bottom of the recess in the holder. Fit the wire as a seat for the window. There should be no overlap at the ends; a small gap is preferable. Previously made indium wire seats for the window should be reusable for at least a few times.

4. Seat the window in its recess and cover it with the top plate of the holder. A second piece of indium wire between the window and the top plate is optional.

5. Insert all the screws round the window holder (with washers if provided) and tighten them progressively. Follow a pattern in which screws diagonally opposite each other are tightened in pairs. Check that the top plate remains

reasonably parallel to the main part of the holder. Continue progressively tightening the screws until they are as tight as can be achieved without damage to the screw heads.

   N.B. This works only with CsBr and CsI; do not attempt this degree of tightening with other window materials.

6. Cut a small piece of indium foil to form a gasket between the window holder and the cold tip of the refrigerator. A rough circle or square with a circular hole at the centre will suffice. Place the indium gasket over the mating thread on the holder. Indium gaskets can be reused, but eventually tend to fragment.

7. Screw the window assembly into the cold tip of the refrigerator until the indium gasket is beginning to be compressed. Make sure that the indium foil is not torn or pulled out of shape in this process. Tighten the assembly by turning the window holder by its edges or shoulders. Tighten it as fully as possible using finger pressure only (both hands may be used). Tidy up the indium gasket by bending any protruding corners into the cold tip. The use of tools, such as a spanner or wrench, for tightening the window holder, though advocated by some, is not recommended. It is important to avoid damaging the thread in the cold tip.

8. If the orientation of the window with respect to the head module is inconvenient—for example, helium lines can foul other components of the system when the window is orientated for matrix deposition or spectral examination—the thickness of the indium gasket can be adjusted or a second gasket added. The author has occasionally used copper or brass washers in place of or in addition to the indium gasket, though this is supposed to be bad practice, because of the lower thermal conductivity of these materials. Whatever adjustments are made, it is important that the window holder is screwed into the cold tip sufficiently tightly. Do not be tempted to adjust window orientation by simply slackening it off. Insufficiently tight screws will result in a high temperature gradient between the cold tip and the window, and this in turn will give poor matrices and excessive evaporation of the matrix host gas.

9. After the window has been cooled at least once to the base temperature of the cold cell, examine the tightness of screws round the window holder and of the mating screw in the cold tip. If any loosening is detected, retighten the screws to the same degree as before. All screws should be checked periodically thereafter but they will seldom need adjustment after the first few cooling cycles.

3.1.2 KBr cold windows
The expense of CsBr and CsI windows has led some workers to experiment with the much cheaper KBr. The main problem is that KBr cleaves very easily on mechanical or thermal shock. It appears, however, that KBr windows can

be rendered more resistant to cleavage by rounding the edges with the aid of a moistened cotton bud, and that such windows can be screwed reasonably tightly into a window holder and thermally recycled many times between room temperature and 10 K without cracking. The author has recently tried a KBr window and found it quite satisfactory, even after a crack appeared as a result of clumsy handling.

3.1.3 Cold windows for UV–visible spectroscopy

CsBr and CsI windows in good optical condition are also usable for UV–visible studies, and if both IR and UV–visible spectra are to be recorded for the same matrices, these would be the substrates of choice. At the shorter wavelengths of UV and visible light, scattering becomes much more of a problem than with IR radiation, so heavily deformed CsBr and CsI windows do not give good UV–visible spectra. Also, CsBr and CsI windows develop colour centres which give rise to absorptions in the UV. The effect can worsen with age. Absorption by colour centres can be removed from matrix UV spectra by background subtraction, but it is still a disadvantage.

For best quality UV–visible spectra, a highly polished CaF_2 window is preferable. CaF_2 is virtually insoluble in water, does not cloud readily, and can take a high polish. It is a brittle material, however, and care needs to be taken when tightening the screws of the window holder. Once the window is properly installed, it should endure many thermal cycles between room temperature and 10 K. Besides their intrinsic high optical quality, polished CaF_2 windows seem also to promote the deposition of matrices of high optical quality, that is, with the minimum of light scattering. It is possible that a smooth substrate surface allows the deposition of more even matrix layers. This effect has been remarked on by numerous workers in the matrix field. CaF_2 is transparent in the IR down to about $840 \, cm^{-1}$, so it is perfectly possible to use a CaF_2 window for combined UV–visible and IR studies, provided that IR frequencies lower than this are not needed.

Sapphire is an expensive alternative to CaF_2 for UV–visible studies, favoured by some workers. It is more durable than CaF_2.

3.1.4 Other window materials

The window materials described above are those which the author has used, but are by no means the only choices. Table 2.1 lists some common window materials for use in the mid-IR, far-IR, UV–visible and vacuum-UV regions. It is easy to experiment with any material if it can be obtained in a size and shape to fit an available window holder, and it is also easy to make special window holders for a particular purpose.

3.1.5 Changing cold windows

If, as is likely, several different windows are to be used in conjunction with a cold cell, it is best to have a separate window holder for each. The complete

Table 2.1. Common window materials and their properties

Material	Useful transmission range		Water solubility (g/100 cm³)	Recommended application
	μm	cm⁻¹		
NaCl	0.25–16	40 000–625	35.8 at 0°C	
KCl	0.3–20	33 000–500	34.7 at 20°C	
KBr	0.23–25	43 000–400	65.6 at 20°C	External windows for IR and UV–visible; economy cold window (IR and UV-visible)
CsBr	0.25–35	40 000–300	124 at 25°C	Cold window (IR and UV-visible); external windows (extended IR range)
CsI	0.25–50	40 000–200	44 at 0°C	Cold window (IR and UV-visible); external windows (extended IR range).
LiF	0.12–7.0	83 000–1450	0.27 at 20°C	Vacuum UV
MgF2	0.11–8.0	91 000–850	0.007 at 18°C	Vacuum UV
CaF2	0.13–12	77 000–850	0.0017 at 26°C	Cold window for UV-visible
BaF2	0.15–13	66 000–800	0.17 at 10°C	
UV fused silica	0.17–2.5	59 000–4000	Insoluble	External windows for UV–visible.
IR fused silica	0.23–3.5	43 000–2900	Insoluble	
Sapphire	0.3–6.0	33 000–1700	Insoluble	Cold window for UV–visible.
Polyethylene	16–2000	625–5	Insoluble	Far IR

window assemblies can then be stored in a safe place when not in use, preferably in a desiccator, and the exchange of windows is facilitated. Removal of a window from its holder always entails a risk of damage and should be done as infrequently as possible.

3.2 Raman and reflection spectroscopy

For spectroscopic techniques not requiring transmission of light through the sample, such as Raman and reflection IR or UV–visible spectroscopy, it is usual to deposit the matrices on to a metal plate screwed into the cold end of the refrigerator (Fig. 2.3(b)). Sample holders of this type can be obtained commercially or else custom made. If necessary, the metal plate can be arranged at an angle to the cold tip to suit particular spectrometer optics. Thermal contact between the sample and the refrigerator heat station is better than with salt windows, since the sample is in direct contact with a metal surface. Gold plating improves the reflectivity in the IR, and good quality reflection–absorption IR spectra can be obtained. One group has developed a multi-faceted cold end which allows several samples to be deposited and analysed separately without the need to warm up the apparatus between samples.[8] The author's group has recently developed a cold cell for diffuse reflectance FT–IR spectroscopy (DRIFTS), which is particularly suited to thick matrices of poor optical quality.[9]

Reflection–absorption matrix IR spectroscopy has also been developed to monitor the eluate from GC columns. A GC–MS–matrix FTIR instrument is commercially available—the Mattson Cryolect. The eluate is diluted with Ar and deposited as a continuous helical matrix on a rotating gold-plated drum, cooled by a closed cycle helium refrigerator.

3.3 ESR spectroscopy

For ESR measurements, the matrix sample must be deposited on a rod which, together with an enclosing tubular vacuum chamber, is narrow enough to be inserted into the cavity of the ESR spectrometer (Fig. 2.3(c)). Both copper and sapphire-tipped copper rods have been employed. It is advisable to use high purity copper. Although the sample set-up is simple enough, there are problems with depositing the matrix on the end of a rod enclosed in the narrow tube. Some special design features of the vacuum shroud are thus required (see Section 4.1).

3.4 Radiation shields

Whatever type of sample holder is installed, it is necessary to shield as much of the cold end of the refrigerator as possible from radiation from the environment. The vacuum system surrounding the sample provides insulation against warming by conduction and convection, but a substantial warming effect is still present owing to incoming radiation. This can be minimized by enclosing the second stage of the expander module within a copper or nickel-plated copper tube attached to the first heat station (see Fig. 2.1). The first heat station is usually threaded externally for this purpose.

A simple radiation shield can be constructed from 35–40 mm diameter copper tube. It is important to make sure that the tube will not touch the second stage of the expander module, or the inside of any shroud to which the head module is to be fitted. The tube can be soldered onto a threaded copper nut to fit the thread on the second heat station. The length of the tube should be such that it reaches down to the top of the sample holder.

Commercial radiation shields are available from the suppliers of refrigeration systems. These are often more elaborate than the simple tube described above, and can enclose the sample holder completely except for two holes to allow matrix deposition and spectroscopic examination. Many matrix workers find these inconvenient in regular use, however. For example, aligning the cold window with the holes in the shield, while still ensuring a tight fit and good thermal contact between the shield and the first heat station, can be difficult unless the shield is constructed in two sections. This type of radiation shield can also impede matrix deposition. Commercial radiation shields therefore commonly end up with their bottom ends sawn off. The extra shielding lost by not enclosing the sample holder as fully as possible—perhaps resulting in a rise in base temperature of 1 or 2 K—does not seem to be very significant in practice.

4. Vacuum shrouds for matrix isolation

With matrix samples at 10 K or thereabouts, it is obvious that they need to be enclosed in a vacuum chamber or shroud. The following are necessary features of any shroud.

(a) The head module of the refrigerator should be mated to the shroud by a rotatable seal—usually by a double O-ring seal—so that the cold window can be rotated within the shroud.

(b) There should be a means for connecting the vacuum-pumping system.

(c) At least one inlet port for deposition of matrices must be provided.

(d) There should be a pair of external windows suitable for the spectroscopic techniques to be employed.

(e) The shroud should fit comfortably into the sample compartment of any spectrometer to be used, such that the matrix can be positioned appropriately and reference beams, if any, are not blocked.

(f) There should be convenient access to the interior of the shroud, when the head module is at room temperature, for cleaning, and so on.

These features, and many optional requirements, can be incorporated into a range of shroud designs.

4.1 Commercial vacuum shrouds

When setting up a new matrix-isolation system, the quickest way to get started is to buy one of the various vacuum shrouds offered by the suppliers of refrigeration systems. There will probably be some customized metal work required, but relatively little. Figure 2.4 shows a typical commercial shroud of compact design. The cold window is enclosed in an approximately cubic stainless steel chamber with an external side of about 6 cm. In Fig. 2.4(a), note the radiation shield enclosing the cold tip of the refrigerator down to a point level with the top of the cold window. This is screwed to the first heat station of the refrigerator and is maintained at about 80 K during matrix experiments.

An advantage of the design shown in Fig. 2.4 is its versatility. There are blanking plates with single O-ring seals on each of the four vertical faces and on the bottom face. Any of these can be replaced by an external window holder or a custom-made plate with a gas inlet. A variety of optical layouts can thus be accommodated while still retaining compactness. The shroud will fit comfortably into the sample compartments of typical spectrometers. Windows and custom made plates must form good seals with the O-rings, which sit in grooves machined in the shroud walls. This requirement determines the maximum size of ports and the minimum size of windows that can be installed on the shroud. For most applications, plates and window holders can be made of any convenient metal, such as stainless steel, brass, copper or aluminium.

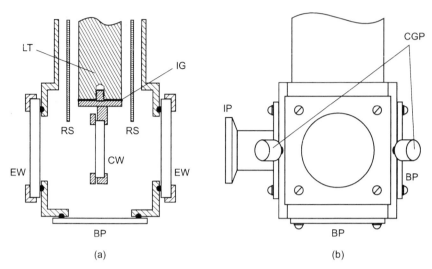

Fig. 2.4 The lower portion of a typical commercial vacuum shroud for matrix isolation. (a) Section through the side view; (b) front view. **BP,** blanking plate; **CGP,** capillary gas inlet port, blanked off; **CW,** cold window; **EW,** external window; **IG,** indium gasket; **IP,** inlet port fitted with NW10 flange; **LT,** low temperature end of the He refrigerator; **RS,** radiation shield.

4.1.1 Inlet ports

The shroud shown in Fig. 2.4 has two capillary gas inlets, placed one on each side of the front window. These are incorporated as standard by the manufacturer and are intended to be used for admission of matrix gases into the shroud. They can be blanked off with suitable plugs, which are also supplied with the shroud. In the experience of many matrix chemists these inlets are too easily blocked by deposits of materials of low volatility, such as the decomposition products of starting compounds. The author, for example, once attempted to deposit a matrix containing diazofluorene through a capillary inlet of this type, and found that the inlet became blocked very quickly. Diazofluorene had been matrix isolated successfully on numerous previous occasions; so it was concluded that either the diazo compound itself or decomposition products (perhaps formed by catalytic interaction with the metal of the inlet) built up inside the narrow bore of the inlet tube and formed a plug. The use of capillary gas inlets can therefore be recommended only for the deposition of mixtures containing volatile and relatively stable components.

For general matrix use a larger diameter inlet port is desirable, and this can be fitted as a replacement to one of the blanking plates. In the shroud shown in Fig. 2.4, the openings have a diameter of 35 mm and the O-rings are 38 mm in diameter. A simple inlet port can be constructed from an NW10 flange and a suitable drilled metal plate (Fig. 2.5). NW10 flanges can be bought in stainless steel, but can also easily be made from brass. The appropriate dimensions are

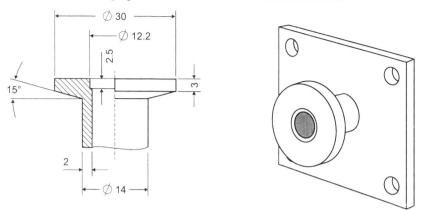

Fig. 2.5 A simple inlet port with an NW10 flange. The profile of the flange is given at left (dimensions in mm).

given in the figure. If brass is used, the flange can be brazed to the plate, eliminating any need for the welding of stainless steel. Screw holes should be drilled to suit the particular shroud.

NW25 or even larger flanges may be substituted if a larger port is required. Dimensions for these can be found in the catalogues of suppliers of vacuum components. Alternatively, metal sockets to fit standard ground-glass cones (e.g. 14/23, 19/26, or 24/29) may be used in place of flanges, thus allowing glass attachments to be made directly to the shroud with the aid of high-vacuum grease or wax. Most matrix laboratories accumulate a selection of different inlet-port arrangements, which can be brought into service when needed.

4.1.2 External windows

The windows fitted to the shroud shown in Fig. 2.4 are disks 45 mm in diameter and 5–6 mm thick. KBr is the most commonly used material, since it is fairly inexpensive and transparent in the mid-infrared (> 4000–400 cm^{-1}) and the UV–visible. For experiments restricted to UV–visible spectroscopy, however, quartz or fused silica windows of about 3 mm thickness might be preferred, because they are more robust. It also is sometimes desirable to fit an extra glass window, to allow inspection of the matrix from a different angle.

If the more extended IR ranges provided by CsBr and CsI are needed, external windows of these materials can be obtained, but they are expensive and more prone to fogging and less easy to polish than KBr.

The external window holder should be a plate about the same thickness as the window, recessed to hold the window and drilled with the appropriate screw holes for attachment to the shroud. An aperture of about 35 mm is suitable for most purposes. If the window holder is not recessed, fitting the window to the shroud is more difficult and the window can slip when the shroud is at atmospheric pressure. Since the window itself forms the seal with the O-ring on the shroud, it is important that the window in its recess should sit proud of the holder by 1 mm or so. Damage to the surface of the window near its rim, such as scratches or chips, can prevent a proper seal being made.

The fitting of windows to a shroud is a regular activity in any matrix laboratory, and requires great care to avoid breakage by uneven or excessive tightening of the retaining screws. The recommended procedure is given in Protocol 2.

Protocol 2.
Fitting an external KBr window to a metal shroud

The following procedure is suitable for fitting windows of all types, but KBr is one of the more easily damaged materials and thus requires the greatest care. The window can be fitted with the shroud in its normal position in the matrix system, provided this allows unimpeded access.

Caution! The KBr window is easily fogged and cracked. Handle it only by the edges and preferably with thin rubber gloves. Put it down only on flat and scrupulously clean surfaces, and avoid any warming or cooling.

Materials
- A suitably sized KBr window
- High-vacuum grease (optional)

1. Carefully remove the O-ring from the groove in the shroud wall, and clean both the O-ring and the groove by wiping with a clean dry tissue.
2. Replace the O-ring in the groove, making sure it is seated properly all the way round. Although not necessary for sealing, a slight smear of high-vacuum grease on the O-ring helps to keep it in place while the window is fitted.
3. Place the window holder on a flat surface and gently lower the window into the recess. The more perfect side, if there is any difference, should be uppermost, since this side will form the seal with the O-ring. Check that the window sits proud of the recess. It may be desired to seat the window in the recess on a bed of soft material such as plasticine or Apiezon Q sealing compound, but this is unnecessary unless the window recess is a little too deep. On no account use bedding material to compensate for an uneven bottom to the recess, and do not use tape, e.g. masking tape, to provide a seat for the window, since it will not compress evenly.
4. Check that the O-ring is still properly located in its groove, and offer up the window in its holder to the O-ring, making sure that it does not fall out of the holder in transit.
5. With one hand, keep the window and its holder firmly but gently in place, then with the other loosely insert all the screws. Without letting go of the

window holder, use a screwdriver to drive the screws in to the point where the window holder can no longer move away from the shroud.

6. Check that the window holder is not skewed, by inspecting the gaps between each edge of the holder and the shroud wall. These should all be the same width and with no perceptible narrowing towards one end. If the holder is not parallel to the shroud wall, adjust it carefully by slackening and tightening the appropriate screws. Slacken first!

7. With the shroud at atmospheric pressure, the screws should be tightened evenly, but with only the slightest possible compression of the O-ring.

8. Evacuate the shroud. If there is a leak around the new O-ring seal, tighten the screws evenly and gently until a good vacuum is achieved, but do not overtighten. If this does not work, there may be a faulty O -ring or defective surface to the window.

9. Once the shroud is evacuated, atmospheric pressure on the window will provide the right degree of O-ring compression. At this stage the screws will have become loose, and should again be tightened evenly but lightly.

10. Avoid any further tightening either under vacuum or when the shroud is at atmospheric pressure. If a leak develops it is more likely to be due to surface defects or a damaged O-ring than to loose retaining screws.

4.1.3 Vacuum ports

When a commercial vacuum shroud such as that shown in Fig. 2.4 is purchased along with a helium refrigerator, a port for attaching the vacuum-pumping system will be provided. A common arrangement has the vacuum port fitted to the head module of the refrigerator. This is inconvenient, however, because rotation of the cold window (e.g. for matrix deposition) is achieved by rotating the head module, and the vacuum port will rotate as well. The vacuum system must therefore be connected by long flexible tubes, reducing pumping efficiency and adding to the general difficulties of using the set-up. Further, in order to gain access to the matrix window for cleaning, and so on, the whole head module must be lifted out of the shroud. This requires disconnection of the pumping system and temporary relocation of the head module to a resting place outside the shroud. Lifting the head module out of the shroud is awkward and requires a fair degree of strength. Long term use of the set-up in these conditions inevitably results in some minor damage, particularly to the delicate thermocouple wires of the head module.

A more convenient arrangement has the shroud in two separable sections, the upper holding the head module of the refrigerator, so that it can be rotated on a double O-ring seal, and the lower constituting the chamber surrounding the cold window. The vacuum port is fitted on the upper section of the shroud. The advantage of this is that the head module of the refrigerator, and hence the cold window, can be rotated without affecting the connection to the

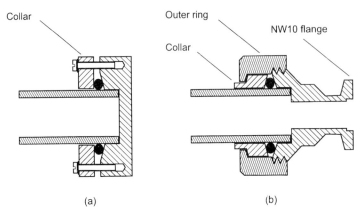

Fig. 2.6 Alternative ways of blanking off a plain-tube vacuum port. (a) A custom-made blanking plate with trapped O-ring seated on a slight bevel. The plate may be square or circular and the O-ring is compressed by a collar fitting around the tube which is tightened by three or four screws. (b) A commercial compression fitting with an NW10 flange. The O-ring is compressed by means of a collar and a threaded outer ring. The NW10 flange may be fitted with a standard blanking plate or an air-admittance valve.

pumping system. Moreover the bottom section of the shroud can be removed, leaving the head module and vacuum connection in place, so that cleaning of the matrix window and the inside of the shroud can be accomplished very conveniently. Figures 2.9 and 2.10 illustrate shrouds of this type, which are discussed more fully below.

Systems with the vacuum port on the head module can be made a little more convenient by blanking off the vacuum port provided—installation of an air-inlet valve is a good way of doing this (see Fig. 2.6)—and relocating the vacuum port to the side of the shroud. This would be the side opposite the inlet port in Fig. 2.4. A vacuum port constructed from brass and copper tube, which is in constant use in the author's laboratory, is shown in Fig. 2.7. Having the vacuum port so close to the cold window invites problems arising from the condensation of diffusion-pump oil or other vapours back-streaming from the pumping system. In practice, however, no such problems have occurred.

4.1.4 Shrouds for ESR spectroscopy

A difficulty with matrix ESR studies is that the cavity of the ESR spectrometer, into which the sample must be inserted, is usually very small. Matrices for ESR study are thus deposited on a rod within a shroud fitted at the bottom with a quartz tubular extension. It is not possible to deposit matrices on the tip of a rod enclosed in a narrow tube, however, since deposition would mostly occur higher up the rod or on the cold end of the refrigerator. It is thus necessary to construct the shroud so that the rod can, in effect, be raised and lowered within it. Figure 2.8 shows one design in which this is accomplished by means of an internal screw thread and O-ring seals. Matrices are deposited

Fig. 2.7 A vacuum port made from standard copper tube and brass. The plate at the lower end fits in place of one of the blanking plates on a commercial shroud, while the top end has an NW25 flange. The length of the copper tube is chosen so that the NW25 flange is high enough to clear the top of any spectrometer that is to be used, but still low enough to allow rotation of the refrigerator head module.

with the tip of the rod within the main chamber of the shroud, while ESR spectra can be recorded with the rod fully inserted into the quartz shroud extension. UV irradiation of the matrices can be carried out through this quartz extension, so the rod needs to be retracted only for matrix deposition. This is just as well, because the task of raising or lowering can be quite laborious and slow if the screw thread on which the lower sections of the shroud move up and down is a fine one.

4.2 Custom-built shrouds

4.2.1 Metal shrouds

Commercially available shrouds are certainly versatile, but for some applications a larger shroud will be needed. Figure 2.9 shows a custom-built stainless steel shroud in use in the author's laboratory. The bottom part of this is a box with external dimensions approximately $10 \times 10 \times 15$ cm. It is fitted with windows in the same way as described for the shroud illustrated in Fig. 2.4. If there is a choice, it is sensible to make windows interchangeable between different shrouds. Two inlet ports, one with an NW25 the other with an NW10 flange, are fitted to the top of the box and angled downwards towards the cold window. The whole of the left side of the shroud is a removable plate seated on an Edwards ISO 63 'Co-Seal' and held in place by four nuts and bolts. This can

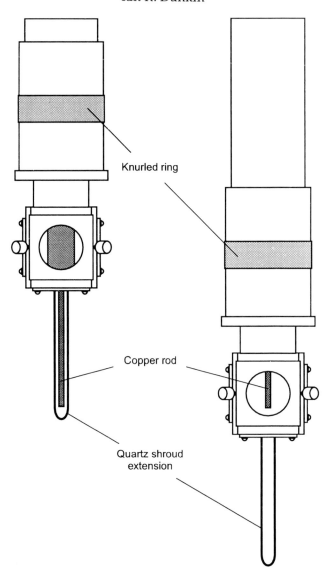

Knurled ring

Copper rod

Quartz shroud
extension

Fig. 2.8 A commercial shroud for matrix ESR studies. The sample holder is a copper rod and the shroud, similar to that shown in Fig. 2.4, has an extension in the form of a quartz tube which can be inserted into the cavity of the ESR spectrometer. The middle and lower sections of the shroud can be raised and lowered on an internal screw and O-ring seals. This is effected by turning the middle section by means of the knurled ring. The shroud is shown in position for ESR measurements (left) and lowered for matrix deposition (right).

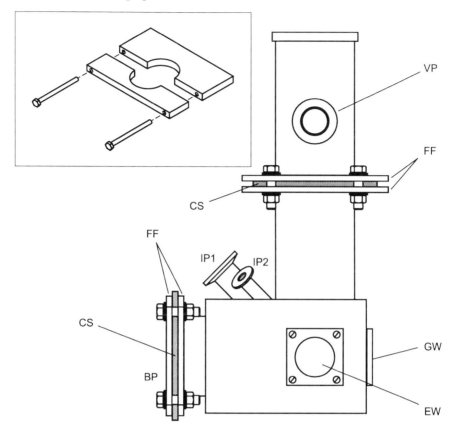

Fig. 2.9 Front view of a stainless steel vacuum shroud used in the author's laboratory. The inset shows a split collar support, on which the lip around the upper section of the shroud can rest. **BP,** blanking plate; **CS,** Edwards ISO 63 'Co-Seal'; **EW,** external window; **FF,** flat flanges compatible with the ISO 63 'Co-Seal'; **GW,** glass window for inspection; **IP1,** inlet port with NW25 flange; **IP2,** inlet port with NW10 flange; **VP,** port for vacuum pump with NW25 flange.

be replaced with other plates holding, for example, further inlet ports or a Knudsen furnace with its electrical feedthroughs and cooling water connections. The opening provided by this arrangement, approximately 7 cm in diameter, and the ample space within the shroud, between the plate and the cold window, provide great versatility. For example, it is possible to deposit simultaneously an organic vapour passing through a hot tube connected to the NW25 inlet port, a beam of Li atoms from a Knudsen furnace mounted on a side-plate, and a stream of argon passing through the NW10 port. Note that in the design of Fig. 2.9 the lower part of the shroud is separable from the upper part, which is fitted with the vacuum-pumping port. The two sections are connected by flat flanges and another ISO 63 'Co-Seal.' A double O-ring seal could be used instead, but the bolted flange arrangement provides support for the heavy bottom end, and is thus to be preferred. The lip round the top of the upper section is useful for supporting the shroud on a split collar arrangement, as shown in the inset in Fig. 2.9.

Although stainless steel is the best material for the fabrication of shrouds, brass and copper can be used if more convenient, or if stainless steel is otherwise inappropriate.

The design of shrouds for matrix ESR spectroscopy poses special problems and at least one group has developed a system in which the ESR cavity is enclosed within the shroud, the whole of which is located between the poles of the magnet.[10,11] This type of shroud must be constructed of a non-magnetic material such as high-grade copper.

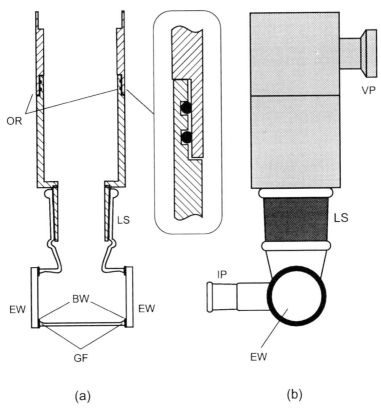

(a)   (b)

Fig. 2.10 A section through the side view (a) and the front view (b) of a glass and brass shroud in three sections used in the author's laboratory. The lower section is made of glass, while the middle and upper sections are of brass. The inset shows a detail of the double O-ring seal between the two brass sections; the O-rings are seated in grooves in the middle section. **BW,** low-melting black wax or Apiezon Q; **EW,** external window; **GF,** ground glass flange; **IP,** inlet port with 24/29 ground glass socket; **LS,** large ground glass socket (60/46); OR, O-ring seal: **VP,** port for vacuum pump.

4.2.2 Glass shrouds

Those with a background in preparative chemistry often feel happier with equipment made from glass. The section of the shroud mating with the head module of the refrigerator should always be made of metal, but the bottom section can be made entirely from glass. Figure 2.10 shows a shroud consisting of brass upper and middle sections and a glass lower section. The glass section is essentially a T-piece made from tubing about 55 mm in diameter, the short tail of the T being fitted with a large ground-glass socket. The brass middle section has an appropriately sized cone and is sealed to the glass section with a black wax, such as Apiezon W40. The two sections can then be removed from the system and replaced as a unit. The two brass sections mate by means of a double O-ring seal.

Many laboratories have quicker turn-round from glass-blowing services than from metal workshops, so having glass sections made with various porting arrangements can be an attractive way of gaining versatility for matrix experiments. Fitting external windows to a glass shroud of this type is something of an art, however. The glass section of the shroud should have each end of the cross-piece of the T thickened and then ground flat to provide a flange with a perimeter about 4–5 mm wide, on which to mount the windows. It is important that the grinding is carried out after any annealing, because annealing can induce a slight curvature in the ground-glass surfaces, and this will result in cracked windows when a vacuum is applied. If the shroud is annealed at some later date, after a repair for example, it is essential that the flanges are reground flat afterwards.

The external windows can be fitted to the ground flanges with a low-melting black wax, such as Apiezon W40, or with Apiezon Q, which is a soft material resembling black plasticine. Wax has a lower vapour pressure, but if wax is to be used, great care must be taken to avoid cracking the windows by excessive thermal shock. The recommended procedure is given in Protocol 3.

Protocol 3.
Fitting external KBr windows to a glass shroud with black wax

The following procedure is suitable for windows of all types, but KBr is one of the more easily damaged materials, and thus requires greatest care. Windows can be fitted equally easily whether the glass section of the shroud is attached to the brass middle section or separated from it, but the appropriate part of the shroud must be removed from the matrix system.

Caution! KBr windows are easily fogged and cracked. Handle them only by the edges and preferably with thin rubber gloves. Put them down only on flat and scrupulously clean surfaces. Only low-melting wax should be employed:

Protocol 3. *Continued*

Apiezon W40, with a quoted vapour pressure at 20°C of $< 10^{-7}$ mbar and a softening point around 40°C, is recommended. Apiezon W, which has a lower vapour pressure but softens at 80–90°C, is death to KBr windows and should be avoided completely.

Equipment
- A metal spatula
- A good hair dryer or gentle heat gun

Materials
- Apiezon W40 wax
- A supply of paper towels and tissues

1. Fit the first window as follows. Place the shroud on a table or smooth surface with one ground-glass flange flat against it and the other facing upwards.

2. Support the glass socket or brass middle section, for example on paper towels or cardboard, so that the lower flange is truly flat on the table.

3. Stuff the inside of the shroud loosely with tissues, to prevent molten wax being blown over the interior.

4. Ensure that the uppermost ground-glass flange and window surfaces are clean.

5. Melt the top of the wax in its can with the heat gun by directing a stream of hot air downwards into the opened can. Spread a coating of molten wax around the flange with a spatula. Make the coat about 1 mm thick. Ensure there are no breaks in the coverage around the flange, but do not worry too much if the coating is slightly uneven. Tidy up any stray filaments of wax with a clean knife.

6. Hold the KBr window in one hand close to the shroud and use the heat gun to warm both the wax coating and the window, using circular motions of the heat gun. Try to avoid blowing filaments of molten wax over the inside and outside of the shroud. Failure to warm the window before placing it on the coating of molten wax can result in cracking. On the other hand, do not overheat the window, as this can result in cracking too.

7. When the wax is warm enough it will appear shiny and smooth. Hold the window by the edges with the forefingers and thumbs of both hands, place it as accurately as possible on the flange, and press down firmly and evenly. The wax should spread and form a visible seal with no breaks.

8. Allow the wax to cool for at least 15 minutes before trying to fit the second window.

9. When the wax on the first window is sufficiently cool, turn the shroud over so that the fitted window is flat against the table. It is best to protect it underneath with a sheet of paper or tissue. It is important to have the fitted

window quite flat on the table, so carefully adjust the support beneath the socket or brass middle section to achieve this.

10. Repeat the waxing operation and window fitting for the second window, not forgetting to warm the window before pressing it on to the molten wax.

11. Allow the wax to cool, then remove the tissue from inside the shroud via the socket. Use a bent wire if necessary.

12. Finally, if the shroud is to be exposed to strong UV or visible radiation such as in matrix photolyses, the black wax should be covered neatly with aluminium foil or in some other way, to prevent excessive warming through absorption of radiation.

13. Allow the wax to cool before returning the shroud to the matrix system and before applying a vacuum.

14. With practice, a very neat wax seal can be achieved. If things go wrong and the shroud will not hold a vacuum owing to poor wax seals around one of the windows, do not try to warm the wax while the shroud is under vacuum. Return the shroud to atmospheric pressure and remove the lower part. It is then sometimes possible to reseat a window by gently and evenly warming it with a heat gun until the wax is softened right round the flange, but great care must be taken since uneven heating can crack the window. It might be better to remove the window (see Protocol 4) and start again, but this too involves the risk of cracking.

15. If a window cracks immediately vacuum is applied, check first for grit between the window and ground-glass flange, then check the flange for curvature. Any curvature must be removed by regrinding.

Protocol 4.
Removing an external KBr window waxed onto a glass shroud

When an external KBr window has to be removed for cleaning or polishing, it may be removed from the shroud after careful melting of the wax. With practice it is possible to remove a KBr window from the shroud in its normal position in the matrix system, but there is less risk if the lower part of the shroud is removed first. Windows which are less prone to cracking than KBr can be removed by the same means, but require fewer precautions.

Caution!   There is even more chance of cracking a window while removing it than while fitting it. Wear thin rubber gloves and apply heat evenly and gently. Use dichloromethane only in an efficient fume cupboard.

Equipment
- A good hair dryer or gentle heat gun
- A shallow vessel, e.g. a Petri dish

Protocol 4. *Continued*

Materials
- A plentiful supply of paper towels or tissues
- Dichloromethane (100–200 ml)   **harmful**

1. Carefully place the lower section of the shroud on a table so that the window to be removed is uppermost and the lower window is flat on the table. Support the shroud with paper towels if necessary.

2. Stuff the inside of the shroud loosely with tissues to prevent molten wax from falling on the lower window.

3. Warm the upper window gently and evenly with a heat gun until the wax appears molten.

4. Apply slight sideways pressure to the window and attempt to slide it off its flange. Do not attempt to pick the window straight off the flange as this will risk putting a strain on it which could crack it, especially if the wax is not quite soft enough. Sliding the window off the flange will tend to spread wax over the surface of the window, but this will be cleaned off in any case.

5. If the window will not move with gentle sideways pressure, give it further gentle warming.

6. Once the window has been removed from the shroud, allow it to return to room temperature, then clean the wax off by immersing it in a Petri dish of dichloromethane and swirling gently. Several changes of solvent may be required.

7. While the window is immersed, it can be wiped with tissues to help remove thicker patches of wax. This is usually a messy process.

8. Once the window is clean, remove it from the solvent and drain it rapidly by holding it vertically.

9. Place the window on a clean tissue and allow it to dry in still air. Do not blow cold or warm air over the window while it is still moist with solvent, because this will increase the rate of solvent evaporation, and the resulting cooling can crack the window.

10. Clean all remaining wax from the ground-glass flange and surroundings with a cloth or tissues moistened with dichloromethane. On no account drop solvent on the lower window, so use solvent sparingly and leave in place the tissues with which the shroud has been stuffed.

11. Remove the tissues from the cleaned shroud.

As the reader will have noted, there is a considerable risk of cracking KBr windows when fitting them with wax or removing them subsequently. For some years, therefore, the author has preferred to fit windows with a sealing compound called Apiezon Q. This resembles black plasticine, remains pliable but firm up to about 30°C, and requires no heating at all. The procedure for fitting a window with Apiezon Q is given in Protocol 5.

Protocol 5.
Fitting an external KBr window with Apiezon Q

This method can be applied equally successfully with any window material.

Caution! The KBr window is easily fogged and cracked. Handle it only by the edges and preferably with thin rubber gloves. Put it down only on flat and scrupulously clean surfaces, and avoid any warming or cooling.

Equipment
- A small knife
- A clean flat surface on which to roll out the Q

Materials
- Apiezon Q sealing compound

1. Place the shroud flat on a smooth surface as described above in Protocol 3. There is no need to stuff the shroud with tissues.

2. On a clean surface, roll out a 'sausage' of Q, about 2–3 mm in diameter and long enough to go right round the ground-glass flange. Make the sausage as even as possible; add more Q if necessary. It is important not to incorporate any grit or other foreign matter into the sausage.

3. Place the sausage of Q on the ground-glass flange and make an unbroken ring by pinching and smoothing the ends together. Pat the ring down with the fingers to ensure that it is adhering to the glass flange.

4. Place the window on the ring of Q and press down firmly at the edges with the forefingers and thumbs of both hands. If the right degree of pressure has been applied the window should adhere firmly to the flange, but test this before moving the shroud.

5. Turn the shroud over and support it so that the fitted window is flat against the table.

6. Fit the second window in the same way as the first.

7. Once the shroud has been installed in the matrix system and vacuum applied, the windows will bed down on their rings of Q, causing a slight extrusion around the edges of the ground flanges, both outside and inside the shroud. On the outside, the excess Q can be trimmed off with a knife or smoothed down to make a neat finish round the window. On the inside, however, it is difficult to trim off excess Q, and attempts to do so are not recommended for fear of scratching the window.

Protocol 5. *Continued*

8. With practice, the amount of Q used to form the seal can be judged to minimize any unsightly spreading inside the shroud. As with black wax, the ring of Q should be covered with foil or other means if the shroud is to be exposed to strong UV or visible radiation.

Windows attached to a shroud by means of Q can be removed, once the shroud has been returned to atmospheric pressure, by carefully cutting round the ring of Q, between the window and the ground flange, with a sharp thin-bladed knife. Any Q adhering to the window can then be removed by gentle rubbing with tissue or a soft cloth. It is rarely necessary to use a solvent.

Apiezon Q has a quoted vapour pressure at 20°C of 10^{-4} mbar, which seems too high for use as a window sealant in a high-vacuum system. In practice, however, no worsening of the attainable vacuum nor contamination of the cold window appear to result from its use. In any case, the vapour pressure will probably be reduced progressively the longer the Q remains in place and exposed to the vacuum of the shroud. The greater convenience of Q compared with black wax and the savings in breakages of KBr windows are un-questionable. If, therefore, it is intended to build a matrix-isolation system with a glass shroud, Apiezon Q is the recommended window sealant.

Glass shrouds are easily broken, and windows fitted to glass shrouds with black wax or Apiezon Q are more vulnerable to damage by knocks than windows mounted on metal shrouds. Great care is needed when the cold cell is moved into a spectrometer sample compartment. Nevertheless, with reason-able care damage can be avoided, and the author has experienced few such breakages during many years of using glass shrouds.

5. Temperature measurement and control

When a closed cycle helium refrigerator is purchased, it is possible to choose from a variety of temperature control units, which will be installed by the supplier. There are three common types of devices for measuring tempera-tures in the range necessary for matrix isolation:

- thermocouples
- silicon diodes
- hydrogen-vapour bulbs.

The last are never used alone.

5.1 Thermocouples

Thermocouples are not the most accurate of the available devices, but their small size, the simple electronics needed for temperature measurement and relatively low cost mean that they are probably the most popular means of measuring temperatures in matrix-isolation work. The most widely used thermocouple consists of a negative thermoelement made of gold containing 0.07 atom-% iron and a positive thermoelement made of a nickel–chromium alloy (Chromel). It has a relatively high thermoelectric sensitivity (> 15 μV K^{-1} above 10 K) and a useful temperature range quoted as 1.4–325 K. Small temperature errors are induced by magnetic fields (at 10 K typically 3% at 2.5 Tesla and 20% at 8 Tesla), but these should not become significant even in ESR matrix experiments.

When installed by the supplier, the thermocouple junction is usually attached to the lower heat station of the refrigerator with indium solder. This is the most convenient location for the junction, because it will remain undisturbed when the cold window is cleaned or exchanged. It must be appreciated, however, that a thermal gradient will exist between the lower heat station where the junction is located and the cold window. To obtain a truer measure of the window temperature it is necessary to drill a small hole in the window itself and press the thermocouple junction into it, perhaps with the aid of a fragment of indium. While it is easy enough to drill holes in CsBr and CsI windows, which are relatively soft, a thermocouple installed in this way is easily pulled out, and it is difficult to ensure that optimum thermal contact with the window is maintained. The thermocouple will also hamper cleaning of the cold window, and will need to be removed when the window is changed.

Few matrix workers put up with the inconvenience of a thermocouple mounted in the window. A more acceptable compromise is to mount the thermocouple junction on the window holder. It can be trapped under a washer by one of the screws around the edge of the holder. The best indication of the window temperature can be expected, if one of the screws furthest from the cold tip is chosen. Mounted in this way the junction is much more likely to remain in good thermal contact with its point of attachment than if it is simply pressed into a hole in the window. It will be less in the way when the window is cleaned, and can easily be removed and replaced using a screwdriver. Of course, it is not good engineering practice to loosen and retighten just one of the screws on the window holder, but at least CsBr and CsI are forgiving in this respect. If a more brittle window is installed it would be safer to loosen and retighten all the screws progressively when removing or replacing the thermocouple.

Many matrix workers are content to leave the thermocouple junction mounted on the lower heat station; the precise temperature of a matrix is seldom critical. Unless there is a compelling reason otherwise, it is probably best to wait until the installed thermocouple is accidentally detached or broken before experimenting with alternative locations for it.

Comparison measurements made in the author's laboratory have indicated that, at base temperature, the window will be 1–4 K above the temperature of the lower heat station, provided that all mounting screws are adequately tight.

Wherever the junction is located, the thermocouple wires will be wound round both the lower and upper heat stations in helical paths with numerous turns. This minimizes temperature gradients in the wires, which could otherwise lead to inaccurate readings. The wires pass through a vacuum seal near the upper end of the refrigerator head module, and are usually connected to an electronic temperature controller. When the head module is removed from its shroud, the thermocouple wires are exposed, and care should be taken at all times not to subject them to any abrasion or other damage.

5.2 Silicon diodes

Silicon diodes are more accurate devices for temperature measurement than thermocouples, but are more expensive to install and more sensitive to magnetic fields. Even miniature versions are bigger than a thermocouple junction. They commonly have operating ranges of 1.4–325 K, though less expensive versions operating at 10–475 K are also available and seem suitable for most matrix-isolation experiments. A great advantage of silicon diodes is their repeatability. Unlike thermocouples they can usually be interchanged without recalibration.

The best place to mount a silicon diode sensor on the cold end of a cryostat is subject to the same considerations as discussed for thermocouples. Because of the size of the sensor, however, it is even less likely that direct mounting on the cold window would be a preferred option.

5.3 Hydrogen-vapour bulbs

A third way of measuring temperatures at the lower heat station is to install a hydrogen-vapour bulb. This is simply a small metal tube attached to the lower heat station and connected by a metal capillary via a vacuum seal to a pressure gauge mounted on the outside of the head module. The whole unit is filled to about four atmospheres with hydrogen gas. When the cold end reaches the boiling point of hydrogen (about 20 K), the pressure gauge will read atmospheric pressure. The pressure inside the hydrogen bulb is very sensitive to temperature over a narrow range each side of 20 K, and this means of measuring the temperature is useless outside the range 16–24 K, but it provides an excellent calibration for thermocouples. Not all suppliers of refrigerators list hydrogen-vapour bulbs as an option, but if a thermocouple is selected as the main means of measuring temperature, the addition of a hydrogen-vapour bulb is recommended. There seems no reason why a custom-made hydrogen bulb could not be installed on a head module provided it has at least one vacant port at its upper end, through which the capillary tube can be routed.

5.4 Temperature controllers and heaters

Not all matrix experiments are carried out at the lowest temperature attainable by the cold cell, and it is often necessary to warm matrices slightly to induce diffusion of trapped species. With closed cycle refrigerators this control of temperature cannot be achieved by controlling the flow of helium from the compressor unit, but is achieved instead by means of a small resistance heater attached to the lower heat station. The usual variety consists of a fine high-resistance wire embedded in a mylar (Melinex) film, and is rated at 20 W maximum power. The heater is connected to the outside of the shroud by wires wrapped round both heat stations and passing through a vacuum seal.

Temperature can be controlled by simply connecting the heater wires to a variable voltage power supply and adjusting the voltage while observing the temperature of the cold tip. It is normal, however, for both the heater and the thermocouple or silicon diode to be connected to an electronic temperature controller, which can be set to maintain any desired temperature. Depending on the specific sensor and controller, temperatures can typically be maintained to within ± 0.1 to ± 0.5 K. The manual method of control can prove temporarily useful if the temperature controller develops faults.

6. The main vacuum system

The shroud enclosing the head module of the refrigerator must be evacuated, and the design of the vacuum system to achieve this is a major consideration in setting up a matrix-isolation system. Books on vacuum technology provide helpful information on building and operating vacuum systems,[12-14] but much may be gleaned from the catalogues of manufacturers of vacuum equipment.

The vacuum serves to insulate the cold sample from warming by conduction and convection (Dewar vacuum) and also to prevent contaminants from being condensed along with or on top of the matrix. Pressures around 10^{-3} mbar are low enough to give an effective Dewar vacuum, but in view of the need to minimize contamination it is important to achieve the highest vacuum practicable. On the other hand, it must be conceded that the sample holder, cold tip of the refrigerator, and the inside of the vacuum chamber of a matrix cell are deliberately exposed to a variety of contaminants in each matrix experiment and that the system also regularly needs to be opened to the atmosphere for cleaning. Ultra-high vacuum (better than 10^{-9} mbar), re-quiring bakeable seals and prolonged pump-down times, is not really practical when it is intended that at least one matrix experiment per day should be carried out. It is therefore usual to settle for a vacuum system capable of maintaining a pressure of about 10^{-5}–10^{-7} mbar inside the sample chamber, when the cold window is at its base temperature. It is sometimes stated that, at this pressure, one monolayer of water per second is laid down on the cold surface of the sample holder or on top of the matrix itself. Since the deposited water does not penetrate far into the matrix, however, this apparent difficulty does not present problems in most matrix studies. The pulsed method of matrix deposition minimizes the incorporation of contaminants, but is not always applicable (see Chapter 4, Section 3.1).

6.1 Pumps

The necessary vacuum can be readily achieved by an oil diffusion pump backed by a two-stage rotary pump. A diffusion pump with an ISO 63 top flange and a nominal pumping speed for air of about $135\,l\,s^{-1}$ will be adequate, but it is better to go for a larger rather than a smaller pump. Diffusion pumps have no moving parts. They require cleaning from time to time and changing of the oil, but maintenance is otherwise confined to occasional replacement of the heater element. In the author's laboratory, one oil diffusion pump has been operating almost continuously for over twenty years, with only one replacement heater element being required in that time.

In a diffusion pump, oil is boiled under reduced pressure, and the resulting vapour passes at high velocity through one or more downward-pointing jets. The rapidly moving oil molecules collide with gas molecules and sweep them downwards, where they are removed from the diffusion pump by the backing pump. In order for the diffusion pump to work satisfactorily a backing pressure of about 10^{-1} mbar is needed. Older diffusion pumps used mercury in place of oil and were usually constructed of glass. Mercury diffusion pumps are not recommended for matrix-isolation systems, because they are fragile and the mercury is easily contaminated. Their use on ancillary vacuum lines, however, poses fewer problems.

Pumping systems, comprising a diffusion pump, backing pump, all necessary electrical switchgear and vacuum interconnections, and possibly one or more vacuum gauges can be bought mounted as a single unit from several manufacturers. Buying the necessary items separately, however, is likely to be cheaper and more flexible.

Turbomolecular pumps will also no doubt be found suitable, though more expensive, alternatives to diffusion pumps. They are cleaner than diffusion pumps because there is no oil vapour within the pump capable of back-streaming into the vacuum system, but they do have a moving part in the form of a rapidly rotating turbo unit, and over long periods will probably require more maintenance than diffusion pumps.

6.1.1 Cold traps and pump protection

Back-streaming of pump oil or other vapours is potentially a source of contamination for matrices, and many laboratories have liquid-nitrogen cooled traps situated between the diffusion pump and the shroud to prevent this happening. In practice, however, back-streaming from modern oil diffusion pumps using appropriate high quality oils does not seem to contaminate matrices to a discernible degree. If the vacuum port on the shroud is high enough above the cold window, back-streaming vapours would in any case condense on the upper part of the cold end, that is, in the area of the first heat station, or on the radiation shield. Moreover, even when the vacuum port is directly in line with the cold window, as is the case when a vacuum port like that illustrated in Fig. 2.7 is fitted, there still appears to be no back-streaming problem.

Another reason commonly given for installing a liquid-nitrogen cooled trap between the cold cell and the diffusion pump is to protect the pumps from contamination by material evaporating when the cold window is warmed up at the end of a matrix experiment. This is an excellent idea in principle, but the truth is that hardly anybody bothers to clean the traps out. They are usually left to warm to room temperature without being isolated from the pumping system, so that the evaporated material ends up passing through the pumps in any case. The expense and extra bulk and inconvenience of a cold trap are therefore hardly justifiable in practice.

6.1.2 Pump connections

Whatever pump configuration is chosen, it will be necessary to connect it to the vacuum port on the shroud of the matrix-isolation cell. It is best if the main pump is fitted with a butterfly valve, either as a separate item or as an integral part of the pump assembly. The pump can than be completely isolated from the shroud and other parts of the vacuum system. Above this valve there should be a vacuum chamber with one or more ports, to which various pipeline connections can be made. The flange on the underside of this chamber should mate with the top flange of the main pump, leaving enough clearance for unimpeded movement of the butterfly valve, assuming that one is fitted. Figure 2.11 shows a manifold design with three ports. One of the ports can be used for connection to the shroud, another for pumping a separate gas line for matrix

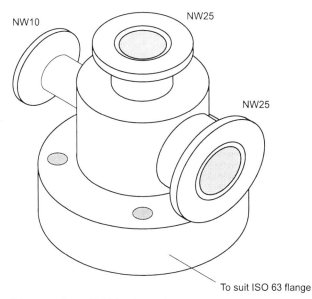

Fig. 2.11 A stainless steel manifold for the main vacuum system of a matrix-isolation unit.

deposition, while the third can be fitted with a vacuum gauge. Spare ports can always be blanked off when not required, so it is best to incorporate at least two or three. Vacuum chambers of this and similar designs have been made from stainless steel and from brass, both types giving trouble free service over long periods.

The exact way in which the vacuum-pumping system should be connected to the shroud will depend on how the cold cell will be mounted. This is discussed in Section 7 of this chapter, which deals with trolleys.

The backing rotary pump should be connected to the diffusion pump via a backing–roughing valve. This allows the diffusion pump to be isolated while the vacuum system is brought to rough vacuum from atmospheric pressure. Once rough vacuum has been attained throughout the system, the rotary pump is returned to its function of backing the diffusion pump, and the system is then pumped down to high vacuum.

If toxic materials, such as metal carbonyls or diazo compounds, are to be used in matrix-isolation studies, safe venting of the shroud via the pumping system must be established. In any case, it is unwise to contaminate the laboratory atmosphere with volatile materials—often of unknown composition—resulting from matrix experiments, however small the quantities. It is thus essential to exhaust the pumping system into an efficient fume cupboard or to the outside of the building. All that is needed is a length of PVC tubing attached to the rotary pump outlet.

6.1.3 Continuous pumping

When the cold cell is at a very low temperature, the best vacuum pump in the system is the cold end of the cryostat itself. In principle, cryopumping by the cold end will draw vapours from a diffusion pump into the shroud and on to the cold window. In practice, as mentioned above, this does not seem to be a problem with modern diffusion pumps and clean, high quality oils. Nevertheless, some matrix workers prefer to isolate the shroud from the pumping system once the cold cell is at base temperature, and certainly after matrix deposition is complete. This technique seems to work satisfactorily, but pumping the shroud continuously seems to work equally well, so which approach to adopt may be simply a matter of taste.

6.1.4 Vacuum failure

Catastrophic failure of the vacuum in the course of a matrix experiment is fortunately a rare occurrence. If it does happen, through a window cracking for example, it is unlikely that the cold window can be saved if it is made of a water soluble material such as CsBr. At first, air will condense on the cold end of the cryostat, and eventually a large amount of water ice will build up. When the cold window finally approaches room temperature, it will turn into an expensive salt solution. It probably will not dissolve completely, but will be irretrievably damaged by pock marks and uneven thinning.

The best policy in these circumstances is to accept the inevitable loss of the cold window and save the rest of the system from contamination. Isolate the pumping system as quickly as possible, for example by closing the butterfly valve. This will prevent large volumes of air and eventually the salt solution from passing into the pumps, an outcome which would require replacement of the oils in both pumps and a great deal of internal cleaning. The refrigerator should also be switched off immediately. A degree of automated protection can be built into the system, as described in Section 6.2.

6.2 Vacuum gauges

6.2.1 Penning and Pirani gauges

Ideally, the pressure inside the shroud and the pressure provided by the backing pump should both be monitored by permanently installed gauges. This combination provides a constant check on the correct functioning of the vacuum system and ease of diagnosing leaks and other faults when they arise.

i. Penning gauges

Penning gauges are robust in construction and easy to clean when they become contaminated. They work in the pressure range $\sim 10^{-4}$–10^{-8} mbar, which is convenient for most matrix-isolation systems. The refrigerator should not normally be operated with the pressure inside the shroud higher than about 10^{-5} mbar, though this limit is often pushed when experimenters are in a hurry, while pressures lower than 10^{-8} mbar are rarely achieved in a matrix-isolation cold cell.

Penning gauge heads operate at 2–3 kV and should be treated with caution when live. Pressure measurement is made by ionization of the gas in the gauge head and measurement of the resulting current. When a Penning gauge is switched on in a vacuum which is already very good, it often takes several minutes for the gauge to give a correct reading. This is quite normal; at very low pressures a cosmic ray is needed to create the necessary ionization of the gas in the gauge head. A small burst of air admitted into the vacuum chamber can also be used to start the gauge, by momentarily raising the pressure to a point where the high voltage in the gauge head will ionize the gas. Penning gauges with digital controllers give no reading or a high pressure warning when the pressure exceeds the their operating range by modest amounts, but annoyingly they cannot distinguish atmospheric pressure from a perfect vacuum. The latter point should not be forgotten if it is intended to use the reading off a Penning gauge for any sort of automatic control of the system (see below).

A Penning gauge to monitor the vacuum in the cold cell should ideally be mounted on a standard T-piece as close to the shroud as possible. It will probably be found easier to mount it near the diffusion pump, for example on a vacuum manifold like that shown in Fig. 2.11, but this will give a less reliable indication of the pressure where it most matters, that is at the cold window.

ii. Pirani gauges

In a correctly functioning vacuum system the backing pressure will typically be 10^{-1}–10^{-2} mbar, and pressures in this range are conveniently measured by a Pirani gauge. This type of gauge measures the effect of the thermal conductivity of the gas on a hot filament inside the gauge head. Such a gauge fitted between the backing pump and the diffusion pump is not strictly essential, but is recommended.

6.2.2 Vacuum monitoring and system protection

A Penning gauge for the shroud and a Pirani gauge for the backing pump can be run from a single electronic controller, with advantages of convenience and compactness. One or more gauge heads can be connected to control boxes, which not only provide the pressure readings but also operate relay switches at pressures which can be preselected. It is thus possible to protect the system to a certain extent by arranging for the refrigerator or pumps to be switched off in the event of a drastic pressure increase. If the cold cell is to be left operating unattended for long periods, such as in overnight photolyses, such protection is essential. The various items of equipment which are to be switched in this way should not be operated directly from the relay switches of the control unit. Apart from the fact that the switched currents may exceed the contact ratings—certainly for the refrigerator—it is important that once the system protection has been triggered, the refrigerator and pumps remain switched off until an operator intervenes. Otherwise it is possible for a matrix experiment to be ruined by an increase in pressure which triggers the protection system, allowing the matrix to evaporate, but for the pumps and refrigerator to be switched back on if the pressure inside the shroud decreases again, after the whole matrix has evaporated for example. Returning to the laboratory to find the cold cell fully operational but the matrix gone can be a puzzling experience, something like happening upon the *Marie Celeste*.

Figure 2.12 shows a double-pole relay circuit which can be used for the safe protection of the system. The live supply for the refrigerator or pump is controlled by one pole of the relay, which is shown inactivated in the figure. The relay is activated by a separate AC supply from neutral (N) and live (L) terminals. The live supply is connected to the relay coil either via a single-pole override switch or via the control unit and the normally open contact of the second pole of the relay. The control unit is set up so that at a satisfactory shroud pressure ($< 10^{-4}$ mbar is suggested), the circuit is maintained, while at pressures above this threshold, the circuit is broken. With the relay inactivated, as shown in Fig. 2.12, and a satisfactory shroud pressure, it is necessary to close the override switch momentarily to activate the relay and thus switch on the refrigerator or pump. With the override switch open again, the relay will remain activated as long as the shroud pressure remains below the preselected threshold. If the pressure rises too far, however, the circuit to the relay coil will be broken at the control unit, and the protected supply will be

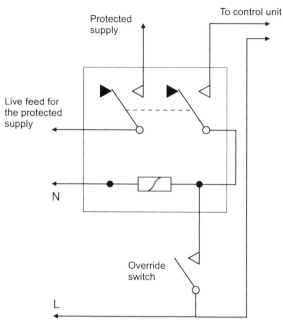

Fig. 2.12 Recommended relay connections to a heavy duty double-pole AC relay for switching off the refrigerator or pumps in the event of a drastic rise in pressure within the shroud.

interrupted. If the pressure falls again to a satisfactory level or if the Penning gauge gives a false reading of very low pressure owing to the shroud reaching atmospheric pressure, the relay will nonetheless remain inactivated until the override switch is closed. More than one protected supply can be switched independently—even at different threshold pressures—by connecting separate relays to appropriate contacts on the control unit. Alternatively, two or three protected supplies can be switched simultaneously by means of a three- or four-pole relay, connected in analogy to Fig. 2.12.

More elaborate protection systems can readily be envisaged. For example, pressure rises in the shroud can be arranged to trigger the closing of electrically operated vacuum valves on the diffusion or backing pumps, rather than simply to switch off the pumps.

If any sort of protection is incorporated into a matrix system which is triggered by a rise in pressure within the shroud, it is essential to override it before any gas is deliberately admitted to the shroud, such as when matrix deposition is to be carried out. With the circuit shown in Fig. 2.12, this is simply accomplished by closing the override switch and leaving it closed until the shroud pressure has fallen again. New users of the system easily forget this precaution, but the frustration occasioned by accidentally switching off the refrigerator ensures that no one makes the mistake more than a few times.

It is possible to protect the system against failure of cooling water supplies to the compressor unit and diffusion pumps, but there is no real need. Interruption of the water supply to the compressor will result in it cutting out, and if this happens without any untoward increase in shroud pressure, it is best if the vacuum pumps remain on. Water-cooled diffusion pumps are very tolerant of low water pressure and will operate for hours with only a trickle or no water at all. Temperature-sensitive (thermal-snap) switches are available as low-cost accessories from most suppliers, and one of these should be fitted to each diffusion pump. This will switch the pump off if the temperature of its external casing rises too far. There is no need to interconnect this safety switch to the rest of the system, however. The shroud pressure will rise once the diffusion pump is off, and the system will in any case shut down as a result of the pressure sensitive protection.

7. Trolleys

For maximum flexibility and ease of use of a matrix-isolation set-up, the cold cell should be mounted on a trolley. This allows the cell to be moved in and out of the sample compartments of spectrometers and, if desired, even between laboratories. There are several common designs, and the best choice in any particular case will depend on the pattern of use intended for the system. For example, if the matrix unit is to be shared among several laboratories, a fully mobile arrangement will be needed, whereas if only one spectrometer is to be used in a single location, a simpler arrangement will suffice, which could even be mounted on a standard bench top. In this section several designs are illustrated, but the possible variations are endless. Design of the best trolley system for a particular laboratory will probably evolve over a period of use.

7.1 Makeshift stands

When a new system is being built, it is natural to begin low-temperature experiments at the earliest possible moment, and in these circumstances a makeshift stand for the cold cell can be made from standard laboratory aluminium rods and clamps. Provided the stand is strong enough and adequate care is taken not to damage any of the equipment, useful work can be carried out, with manual lifting of the head module in and out of the spectrometer. Such temporary arrangements are to be recommended initially, because they allow the refrigeration system to be tested as early as possible after delivery and give the users of the system a clear idea of how the various components are fitted together. Preliminary experience of this sort, even if it is only running a few spectra of stable compounds in matrices, can be a valuable aid to designing a more permanent set-up.

7.2 Floor-mounted trolleys

The most usual type of trolley for matrix isolation runs on the laboratory floor. At an early stage it must be decided if the cold cell is to be mounted vertically (see Fig. 2.13) or horizontally (Fig. 2.14). With the former arrangement the cold window can be rotated about a vertical axis by rotating the refrigerator head module, while in the latter case the cold window rotation takes place about a horizontal axis. It is probably slightly easier to construct a trolley with a horizontally mounted cold cell, but this arrangement has the disadvantage that, in most cases, the gas lines for matrix deposition must enter the shroud from above (see Fig. 2.14). This position of the spray-on line is not very convenient for the sublimation of solids or liquids of low volatility, where a line-of-sight path from sample to cold window is sometimes necessary. It is also rather easy to drop small objects into the shroud when connections to the inlet port are exchanged or modified. For these reasons vertical mounting of the cold cell will generally be preferred, but either set-up can be used successfully.

7.2.1 Wheels and manoeuvrability

The trolley should be fitted with sturdy wheels without castor action, so that

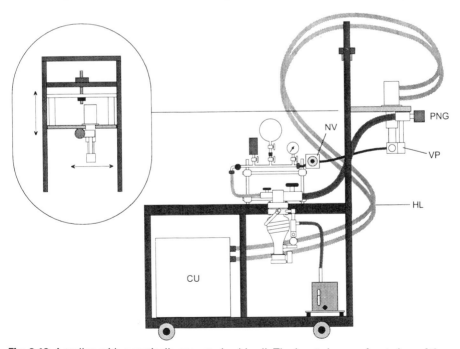

Fig. 2.13 A trolley with a vertically mounted cold cell. The inset shows a front view of the mounting arrangements of the cold cell, which allow vertical and left–right movement. **CU**, compressor unit; **HL**, helium lines; **NV**, needle valve; **PNG**, Penning gauge head; **VP**, vacuum port (cf. Fig. 2.7).

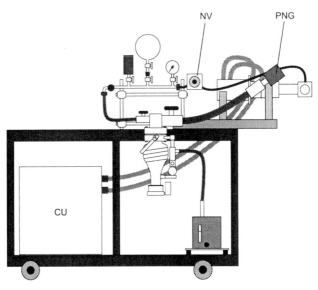

Fig. 2.14 A trolley with a horizontally mounted cold cell. **CU,** compressor unit; **NV,** needle valve; **PNG,** Penning gauge head.

the cell can be wheeled in and out of the spectrometer sample compartment in an accurate straight line. The accuracy and repeatability of this motion can be improved by running the trolley on or between rails fixed to the laboratory floor. As well as this back-and-forward movement, it is best to be able to move the cold cell by small amounts in the vertical direction and, if possible, to left and right, so that optimum positioning of the cold window in the spectrometer sample beam can be easily achieved. A provision for vertical and horizontal movement of about 10 cm in each direction should be ample. It is a good idea to fit adjustable stops, such as on the rails or on the floor, so that once the optimum position is determined the cold cell can be returned to it rapidly. A brake attached to one of the wheels, or arranged to clamp on to a rail or the floor is also helpful in maintaining the position of the cold cell.

With trolleys of the type shown in Figs 2.13 and 2.14 the cold cell can be moved between two or more spectrometers by jacking up and manoeuvring it, or by having the spectrometers also mounted on trolleys. If enough space is available, the ideal arrangement is to have each spectrometer on its own trolley running on rails at right angles to the direction of motion of the cold cell trolley or on castors. Remember that spectrometers are delicate instruments, however, and take care not to shunt them about too vigorously. Adjust the sample beam heights of all spectrometers to be as nearly as possible the same. This reduces the time taken to adjust the cold cell when changing from one to another. Spectrometers with low beam heights, such as many UV–visible spectrometers, can be placed on substantial wooden blocks or metal supports, provided these are strong enough and there is no danger of the instrument slipping off. It is perfectly straightforward to move the cold cell between spectrometers with all the systems operating, and thus to examine a matrix by more than one spectroscopic technique in a single experiment.

If it is necessary to move the trolley further away, for example to another laboratory, it can be manoeuvred by jacking up the end away from the head module on a standard trolley jack, which then allows swivelling of the whole unit through fairly tight turns. A strong jacking point should therefore be incorporated at the mid point of this end of the trolley. Alternatively, two additional castor wheels on strong steel screws can be fitted to this end. These can be either screwed down to take the weight of the trolley, in which case it can be manoeuvred easily, or screwed up, allowing only back-and-forward motion on the non-castor wheels. For ease of manoeuvring, the trolley needs to be well clear of any rails on the floor, so, if it is intended that the trolley should be moved between two or more locations, rails should be dimensioned and installed to allow enough space for the trolley to be wheeled clear. It is unlikely to be necessary to move the trolley between laboratories while the cold cell is at low temperature, though for laboratories on the same storey of a building this could be achieved with some ingenuity in maintaining electrical and water connections. If movement to other laboratories is intended, check that the whole trolley assembly can pass through any doors that intervene. If the system is to be moved between floors of a building, check that it fits into the lift.

7.2.2 Locating the compressor unit

It is not necessary that the compressor unit be accommodated on the trolley, but doing so has two advantages. Firstly it makes the whole unit self contained, thus increasing the convenience of wheeling the system about. Secondly, the weight of the compressor unit provides a good counterbalance to the head module, which will normally overhang the end of the trolley by a considerable distance. If the cold cell is to be used with only one spectrometer, however, a much simpler arrangement will suffice, and the compressor unit can be placed under a bench or in any other convenient location.

7.2.3 Locating the main vacuum system

The major components of the vacuum system are normally fixed to the main part of the trolley, that is, not to the adjustable part which holds the head module. The vacuum system thus needs to be connected to the shroud by suitably flexible coupling such as a flexible stainless-steel pipeline. Pumping speed and efficiency are improved if this pipeline is made as short and as wide bore as possible, but the need for flexibility places limits on the practicable dimensions. Flexible stainless steel pipelines fitted with standard flanges are commercially available in various sizes. A 1 m length of stainless-steel pipeline with an internal bore of about 41 mm and fitted at each end with an NW40 flange is a good compromise. Rotary pumps are best mounted low down on the trolley with rubber feet to minimize vibration, and connected to diffusion pumps by flexible tubes, which can be of stainless steel, thick walled rubber or any of a variety of other materials.

The lower part of a diffusion pump is very hot when the pump is working and should be guarded by some means to prevent anyone using the equipment from accidentally touching it. A wire gauze cage around the lower part serves well and still allows air to circulate.

In Figs 2.13 and 2.14, note that the Penning gauge head is located quite near to the vacuum shroud, and not on the chamber immediately above the diffusion pump. Note also that these diagrams do not show some of the ancillary equipment such as control boxes for the vacuum gauges and electrical and water services to the trolley. The placement of these is not usually critical and the most convenient layout can be determined by experience.

7.2.4 Spray-on lines

A small vacuum line for depositing, or *spraying on*, matrix gases is needed on the trolley. It can be made of glass or metal, but it is obviously easier to see that it remains clean if it is a glass line, whereas metal lines are more robust. A reduced version of the preparative vacuum line described in Chapter 3 would be satisfactory. Metal lines can be made from stainless-steel tubing and compression fittings. The spray-on line should be fitted with a Bourdon or barometer gauge reading in the range 1–1000 mbar to monitor the pressure of the host gas, a sensitive Pirani or Penning gauge to monitor high vacuum, a valved port for connection of gas bulbs, and a needle valve to control the rate of spray-on. The needle valve should be connected to an inlet port on the shroud by means of a flexible pipeline. A 0.5 m length of pipeline with a bore of about 12 mm and fitted with NW10 flanges should suffice for vertically mounted cold cells, although a longer pipeline may be needed for horizontal cells. More complicated vacuum lines may be necessary for special applications, for example when two gases are to be deposited simultaneously.

Whatever the configuration of the spray-on line, it can either be provided with its own pumping system, such as a small diffusion pump and rotary backing pump, or it can be pumped by the main vacuum system. In the latter case, the head of the main pump should be a manifold chamber with at least two ports (cf. Fig. 2.11), two of which are fitted with valves. The shroud is then connected via one of these valves and the spray-on line via the other. Some care needs to be exercised when using this set-up. During an experiment with the cold window at low temperature, the valve between the pump and the shroud should always be fully closed when the spray-on line is being evacuated, and it should not be reopened until the spray-on line has again been isolated and the high vacuum restored in the manifold. With experience, however, there should be few problems in using just one pumping system for both purposes, and the saving in cost and complexity is worthwhile. Some matrix workers prefer, in any case, to keep the main pumping system isolated from the shroud once the base temperature of the cold cell has been attained, and in these circumstances it is available most of the time for manipulations on the spray-on line.

More details about the design and operation of the spray-on line are to be found in Chapter 4 (Section 3).

7.3 Gantry-mounted trolleys

Floor-mounted trolleys as shown in Figs 2.13 and 2.14 are the most common and most adaptable, but if floor space is very limited, it is possible to remove all of the major components of the matrix-isolation system to a trolley on an overhead gantry. Figure 2.15 shows such a set-up, which was in use in the author's laboratory until a few years ago. Both the head module with its shroud and the diffusion pump were suspended from a central steel rod of about 25 mm diameter. The shroud and pump could thus be rigidly connected by relatively short lengths of 25 mm diameter copper tube. The cold cell and its diffusion pump could be moved in all three dimensions: the central rod was threaded at its upper end and could be screwed up and down by means of a large knurled nut; this rod was mounted on a cross-member which could slide in and out of the spectrometer sample compartment on two aluminium cylinders fitted with linear bearings running on two steel rods (see inset to Fig. 2.15); and the whole was suspended from a trolley running on a substantial gantry made of U-section mild steel resting on a standard laboratory bench. The rear of the gantry was bolted to the laboratory wall, giving a rigid structure. The compressor unit was mounted on the gantry trolley as also was the spray-on line (fitted in this case with its own pumping system). Only the two backing rotary pumps were placed on the floor.

As Fig. 2.15 illustrates, the cold cell could be moved between an IR spectrometer and a UV–visible spectrometer with the greatest convenience. It was necessary merely to pull the cold cell forwards out of the sample compartment of one spectrometer, roll the trolley along the gantry to the second location, and push the cold cell gently back into the sample compartment of the second spectrometer. In the forward position, the cold cell was conveniently located for matrix photolysis. The disadvantage of the set-up, however, was its lack of flexibility. The matrix-isolation system was literally bolted to the wall, and could not be used in any other location. Moreover, the compressor unit was very heavy, and installing it on the gantry and removing it for servicing were difficult and hazardous operations. It would not have been necessary to mount the compressor on the trolley, however, if more space had been available. With this alteration, the gantry arrangement can be recommended for matrix systems in which no more than one or two spectrometers are to be used, and where there is no foreseeable need to move the unit to another laboratory.

The set-up shown in Fig. 2.15 had a brass and glass shroud of the type shown

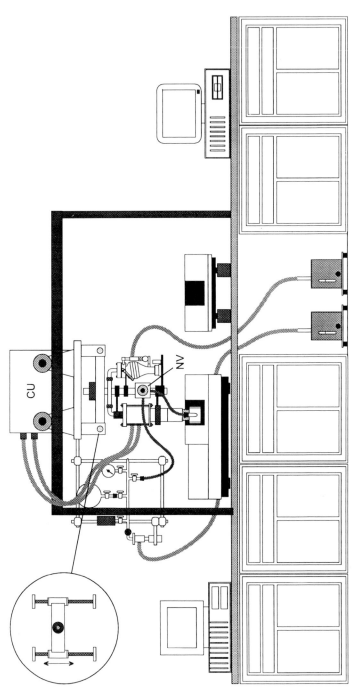

Fig. 2.15 A matrix-isolation unit mounted on a gantry. The compressor unit, head module, vacuum systems, cold window and shroud are mounted on an overhead gantry above a standard laboratory bench. The matrix sample can be moved very easily between an IR spectrometer (left) and a UV–visible spectrometer (right), or brought forward for photolysis. The inset shows a top view of the sliding cross-member on which the cold cell and diffusion pump are mounted. **CU**, compressor unit; **NV**, needle valve.

in Fig. 2.10. Note that the needle valve used to control matrix deposition was mounted on the same rod that supported the cold cell and diffusion pump, and was connected to the inlet port of the shroud by a flexible pipeline. This was adjusted to minimize any strain on the glass shroud, and then remained strain free even when the cold cell was moved. The needle valve was connected to the glass spray-on line by a longer flexible pipeline, which had to accommodate the movement of the cold cell in and out of the spectrometers. The pipeline was clamped firmly close to its connection to the spray-on line and was of a generous length, so that movement of the cold cell did not impart any strain to the spray-on line. If accidents are to happen, this is the best arrangement, since it is less traumatic to fracture the spray-on line than to fracture the shroud, especially when the cold cell is at low temperature.

7.4 Electrical and water supplies for trolleys

A matrix-isolation system requires both electrical and cooling water supplies. The diagrams of trolleys (Figs 2.13, 21.4, and 2.15) show the major components of the systems, but they do not show some of the electronic components, such as vacuum-gauge controllers and temperature controllers, or the main electrical and water supplies.

7.4.1 Electrical supplies

Mains electricity is required for the refrigerator, vacuum pumps, vacuum gauges, temperature controller, and any additional electrical equipment, such as pyrolysis tubes or Knudsen furnaces. The total load can easily exceed the 3 kW limit of a single fused 13 A supply, which is standard in the U.K. It is therefore best to serve the trolley with two 13 A supplies, one to power the refrigerator, and the other feeding a trailing socket with six switched outlets. All the other electrical equipment can be run off the trailing socket, which should be screwed or bolted firmly in a convenient location on the trolley.

Elaborate control panels, which can be designed with tidy connections of the equipment all out of sight, neat switchgear, and with fuse holders and vacuum gauges mounted on the surface, are attractive until equipment failure produces the need to interchange components. Experience teaches that having each electrical component fitted with its own standard plug (i.e. a 13 A plug in the U.K.) increases flexibility and eases life in a crisis.

When fully functional, a matrix-isolation trolley has a good deal of electrical cable festooned over it. Make sure that all cable is kept away from hot surfaces such as the bottom part of a diffusion pump.

The cables leading from the trolley to the electrical supply points should be of adequate current rating, and should be protected from damage by abrasion, heat or simply being walked on. Routing the cables overhead is usually the best option. The cables should be long enough to allow movement of the trolley within the laboratory, to the fullest extent needed, without interruption of the supply. European 16 A plugs and sockets, which have a positive engagement to prevent accidental disconnection, are a good choice for in-line connectors.

7.4.2 Cooling water

Unless air-cooled models are chosen, cooling water will be required by the refrigerator compressor unit, the diffusion pump, and some other ancillary equipment such as pyrolysis tubes and Knudsen furnaces. This means that the trolley has to be equipped with at least two hoses: one to supply the water and the other to return it to a drain.

It is convenient to carry out some elementary plumbing and make two manifolds from standard copper tube, soldered or compression-fit T-pieces, and the sort of taps used on the water supplies of domestic washing machines and dishwashers. One manifold is connected via a stopcock and reinforced hose to a water supply, and the other via a reinforced hose to a drain. A stopcock is optional on the second manifold. A pair of taps, one on each manifold, can be used as the water feed and drain for the compressor unit, another pair for the diffusion pump and further pairs for ancillaries. Spare taps are turned off, but can be further protected from leaks by interconnecting feed and drain pairs by a short length of nylon-reinforced hose. All hoses should be of high quality and reinforced with nylon braid, and all connections should be made with jubilee clips or other means of ensuring a leak-free joint even when the water pressure increases.

Hoses for the main water feed and drain should be of sufficient bore to provide the necessary supply at the available pressure, but standard 12.5 mm hose is likely to be adequate. Good quality reinforced garden hose or nylon reinforced PVC should prove satisfactory, provided that steps are taken to protect it from damage. An overhead route from the trolley to the supply point and drain will minimize the risks. As with the electrical cables, the hoses should be long enough to allow free movement of the trolley.

8. Care and maintenance of a matrix-isolation system

Looking after a matrix system once it has been set up is largely a matter of common sense and following manufacturers' recommended procedures for operating and servicing equipment. Nevertheless, the average system is fairly complex, and all of the major components have to function properly if useful results are to be obtained. In this section a few pieces of advice, which have been learned by experience, are passed on.

8.1 The refrigerator
8.1.1 Helium pressure

Fortunately, the compressor and head modules of a closed cycle refrigerator require very infrequent servicing. Topping up with helium is needed from time to time, but if this becomes necessary more often than about once every three to six months, efforts to find and cure the leak should be made. New refrigerators commonly operate for two or three years before the helium needs to be topped up. Frequent uncoupling and recoupling of the helium lines between the compressor unit and the head module will eventually result in a significant loss of helium and should be avoided. Topping up with helium requires a helium cylinder fitted with an appropriate valve or regulator and a length of 6 mm o.d. copper tube with a compression fitting at the equipment end. The way in which helium is introduced to the compressor unit varies between models, and the instructions provided by the manufacturer should be followed faithfully. It is important to avoid air contamination of the helium.

Compressor units are fitted with a cut-out switch which operates if the helium pressure falls below a certain minimum. The switch is reset by a button, and it is sometimes possible to coax a reluctant compressor unit into life by holding this in for a few seconds when the unit is switched on, thus allowing the pressure to build up in the helium-feed line. Such practices should be used only in an emergency; and the helium should be properly topped up at the earliest possible moment. Most compressor units are fitted with a helium-pressure gauge, so there ought to be no excuse for allowing the helium pressure to fall below the permitted minimum, unless a leak develops.

Over long periods, or if contamination of the helium has been allowed to occur through carelessness, charcoal filters within the compressor unit may need to be regenerated or replaced. It is usual to return the compressor to the suppliers for this procedure, but it may be possible to regenerate a filter unit in the user's own laboratory, and advice should be sought from the suppliers if this is contemplated.

8.1.2 Helium lines

Each of the helium lines between the compressor unit and the head module consists of a flexible corrugated metal inner tube surrounded by a strong metal braid. The ends are fitted with sprung screw fittings of a design which allows coupling and uncoupling to be carried out without contamination by air. As the coupling is screwed up, a small burst of helium is released just before the seal is made, displacing air from the coupling. A small loss of helium is thus incurred each time the helium line is recoupled. The helium lines are maintained under high pressure (15–22 bar) and the external metal braid is there to stop the corrugated tubing expanding and bursting, not merely to protect the tube from abrasion. Newer refrigerators are fitted with stainless-steel helium lines, but older ones can be found with copper lines.

In use, the helium lines need to be looked after at all times, and the route which they take from the compressor unit to the head module is critical. When the head module is rotated the helium lines need to move as well, and unless they are routed carefully, they can impose a severe drag on the rotation. In addition, the external metal braid on the lines can take the finish off spectrometer cases and other pieces of equipment if it is allowed to rub against them. The lines should be taken through gentle curves rather than tight turns and should never be used as handles for rotating the head module. Unless care is taken at all times, the helium lines will eventually fracture.

Fractured helium lines are expensive to replace but can be repaired. The older copper lines are the easiest to repair, requiring only brazing. The most common type of fracture occurs near to one end of the corrugated inner tube. The fractured section can be cut out, shortening the line slightly, and the new end brazed back into the terminal copper fitting. It is important to restore the external braid to give full support to the inner tube, so this also should be brazed to the terminal fitting. Workshops carrying out this repair should be instructed to avoid allowing flux or other contaminants to enter the inside of the helium line. Repairs to stainless-steel lines are more difficult, requiring welding, but the same principles apply. Since fractures usually occur to the helium lines when they are in use, the whole compressor unit and head module, as well as the repaired tube, will normally need to be refilled with helium when the repair to the line is complete. Instructions for this operation will be found in the operator's manual for the refrigerator and should be adhered to.

Helium is an exceptionally fluid gas, and all materials are porous to it to some extent. The metal helium lines provided with the refrigerator are the best from the point of view of containing the helium and maintaining its pressure. Nevertheless, these lines do have some disadvantages, amongst which are relative lack of flexibility, abrasive exteriors, ease of fracture and weight. When a helium line fractures, one can add the further disadvantage of high replacement cost. Although manufacturers and suppliers of refrigerators throw up their hands in horror at the suggestion, the helium lines can be replaced with much cheaper, lighter, more flexible and less abrasive alternatives.

The coiled lines used in air brakes are particularly light and flexible, and the author has seen them incorporated into a matrix-isolation system designed to be used in magnetic circular dichroism (MCD) experiments. They are available with a male ¼″ BSP (British Standard Pipe) fitting at each end, and with suitable adapters can be mated to the terminal fittings of a metal helium line, once these have been cut off. The originator of this idea stated that the helium pressure decreases faster with these lines than with metal lines, but is maintained within the correct range for months, so that topping up with helium does not need to be carried out excessively frequently.

The author has recently experimented with a hydraulic hose as a replacement for a fractured helium line. The hose is made of nitrile rubber with two reinforcing braids of high tensile steel wire and a synthetic rubber outer cover. Although not so light and flexible as the coiled air hose, it is much more flexible and much less abrasive than a braided metal helium line. It is rated to withstand pressures of about 300 bar, exceeding by an order of magnitude the minimum requirements for a helium line. The hydraulic hose can be supplied with a variety of fittings, to which it is mated with the aid of a tiny amount of lubricating oil, and which can be chosen to fit the terminal connections of a metal helium line, once these have been cut off. The replacement line has been in use for over a year, and the rate of decrease in helium pressure is perfectly tolerable, topping up being necessary every three to five months. It is far more convenient in use than the original braided metal helium line, but the author so far lacks the courage to saw the ends off a perfectly good metal line in order to replace the second helium line on the system. The cost of the hydraulic hose and its fittings is a small fraction of the cost of a new metal line; so at least the fear of fracturing the second line has been much diminished.

8.1.3 Refrigerator failures

Excessive vibration or loss of cooling capacity indicate a problem in the refrigerator, most often within the head unit. With advice from the suppliers and following any relevant instructions in the operator's manual, it is possible to disassemble the head unit and effect some simple remedies without returning the module to the suppliers. The author has successfully replaced a valve motor in a head module in this way, and cleaning the regenerators, which can become necessary, is supposed to be feasible (see Fig. 2.1). Note that the O-rings in the head module are of a specially hard type and should not be replaced with softer ones.

Causes of failure other than those already mentioned will probably require all or part of the refrigerator to be returned to the supplier. Fortunately closed cycle refrigerators tend to be very robust and can be expected to give many thousands of hours of trouble free service.

8.2 The shroud and vacuum system

8.2.1 Maintaining low pressure

If possible the shroud and vacuum system should be kept under high vacuum at all times, closing down only for periods of disuse of a week or more. Even when the vacuum system is given a holiday, all valves should be shut, and the interior pressure allowed to rise as slowly as possible. Once the system has reached atmospheric pressure, it takes an appreciable time to achieve the lowest pressure again. In particular, atmospheric water is adsorbed on interior surfaces and can be difficult to remove.

Modern pumps are very robust and can be left running continuously with little attention except to ensure that oil levels and water supplies are maintained. The water pressure in laboratory buildings often rises considerably at night and at weekends when a lot of equipment is turned off, so it is important to guard against leaks due to inadequate water connections or burst hoses. Water supplies should run only in reinforced hoses in good condition, and all water connections should be made with properly tightened jubilee clips or something similar.

Avoid allowing large volumes of air or other gases to pass through the diffusion pump, since this will blow oil out of the pump. Diffusion pumps work quite well on much less oil than their nominal charge volume, but eventually there will be insufficient oil to give an acceptable vacuum. Organic and other vapours passing through the diffusion pump will tend to decompose and build up a carbonaceous residue in the pump. The residue will eventually reduce the efficiency of the pump to a significant degree and will then need to be removed. This entails dismantling the pump, withdrawing the jet assembly and cleaning both the interior of the pump and the jet assembly with solvent and possibly even mild abrasives. The interior surfaces at the bottom of a diffusion pump are not very accessible, so recourse to long screwdrivers or bent wire is often necessary. The quantities of material involved in matrix experiments are usually very small, so pump cleaning need not take place very frequently if reasonable care is taken. Once every year or two would be normal.

When condensable materials are allowed into the rotary pump, they tend to contaminate the oil, giving it a significant vapour pressure. Even a small amount can increase the backing pressure to the point where the diffusion pump will no longer work. Do not attempt to operate the diffusion pump with a high backing pressure, since overheating and deterioration of the diffusion-pump oil is likely to result. A contaminated backing pump can usually be cleaned by purging with a controlled stream of air for several hours (i.e. ballasting). Most rotary pumps have a purge valve for this purpose.

When the vacuum in the main system is poor, it should be remembered that this might be due either to a leak or to out-gassing of material which has been adsorbed on interior surfaces. If the latter is the cause of poor vacuum, pass warm air from a moderate heat gun over the metal parts of the system, while monitoring the system pressure. A section which yields a sharp rise in pressure when warmed should be heated until the pressure drops to a constant value. Do not overdo the heating, though, since elastomer seals can be softened and deformed by excessive heat, and on no account pass hot air anywhere near a KBr window. Heating tape can be wrapped around the outside of the affected section of the shroud or pumping system, which can then be subjected to prolonged heating. This is often necessary for 'sticky' absorbates such as Br_2 and I_2, but is advisable only for metal shrouds. Some types of heating tape are unsuitable for metal surfaces; so care should be taken in selecting tape for this purpose.

Air leaks need to be located before they can be cured. Spark testers are not much use with vacuum systems made largely from metal components. A simple way of detecting leaks is to spray droplets of dichloromethane on to a portion of the vacuum system, while monitoring the system pressure. Penning gauges respond well to the ingress of small amounts of dichloromethane; the pressure reading can go down as well as up. Any response indicates a leak in the area being probed. Avoid spraying solvent near a KBr window. The cooling effect of solvent evaporating can crack it even in normal conditions, but if there is a leak between the window and its seat, solvent will be sucked into the shroud and instantly vaporized with a drastic cooling effect. Cracking of the window will then be inevitable.

Once a leak has been located the cure is usually obvious. In an emergency, leaks can be sealed temporarily with Apiezon Q, applied to the outside of the defective part of the system. The Q should be pressed down firmly to make a good seal. Leaks should be properly repaired as soon as possible, however. A vacuum system encrusted with blobs of Q is as unreliable as it is unsightly.

8.2.2 External windows

External windows made of materials which are hygroscopic and prone to fogging, such as KBr and CsBr, may need to be protected from atmospheric moisture when the cold cell is not in use. The usual way of doing this is to select a medium size transparent plastic bag with no holes in it and cover the bottom with self-indicating silica gel. The mouth of the bag can then be taped around the shroud to enclose the part where the windows are mounted and form the best possible seal. In some laboratories this procedure is followed routinely, but it is somewhat inconvenient. In the author's laboratory in Glasgow, a place noted for its damp climate, external KBr windows are always left unprotected, and no undue fogging has occurred during many years. When the more hygroscopic and much more expensive CsBr windows are installed on the shroud, however, a silica gel bag is fitted.

Over a period, external windows sometimes build up deposits on the inside from evaporating matrices. Sometimes these are stubborn polymers. Windows in this condition must be removed for cleaning or replacement. Procedures for fitting and removing external windows have been given in Section 4 of this chapter.

8.2.3 The cold window

Cold windows usually need cleaning after each matrix experiment, but seldom need to be removed from the cold tip for this purpose. Unless a specially tenacious residue has been formed, it is best to wash the window and its holder with a solvent such as dichloromethane, sprayed on from a plastic wash bottle. The window and holder can be wiped with a clean soft cloth or clean tissue if necessary. The cloths used for cleaning lenses and other optical components are best, but they should be washed or replaced regularly, otherwise they quickly become contaminated all over. Residues on the radiation shield and lower heat station can be removed at the same time if they are present. Try to avoid spraying solvent over the heater and thermocouple or diode wires, or the electrical insulation may be destroyed. Dry the cleaned parts with warm air from a hair dryer or heat gun, but avoid excessive heat, which could melt the insulation around wires.

Because the cold window is kept in the shroud, which should always be under reduced pressure, there is no need to protect it in any other way.

References

1. Bass, A. M.; Broida, H. P., eds. *Formation and Trapping of Free Radicals.* Academic Press; New York, **1960**.
2. Meyer, B. *Low Temperature Spectroscopy*; Elsevier: New York, **1971**.
3. Hallam, H. E., ed. *Vibrational Spectroscopy of Trapped Species*; Wiley: London, **1973**.
4. Cradock, S.; Hinchcliffe, A. J. *Matrix Isolation*; Cambridge University Press, **1975**.
5. Moskovits, M.; Ozin, G., eds. *Cryochemistry*; Wiley: New York, **1976**.
6. Andrews, L.; Moskovits, M., eds. *Chemistry and Physics of Matrix-Isolated Species*; North-Holland: Amsterdam, **1989**.
7. See, for example, Rose-Innes, A. C. *Low Temperature Laboratory Techniques*, 2nd edn;English Universities Press: Liverpool, **1973**.
8. Hauge, R. H.; Fredin, L.; Kafafi, Z.; Margrave, J. L. *Appl. Spectrosc.* **1986**, *40*, 588–595.
9. Dunkin, I. R.; Gallivan, S. L. *Vib. Spectrosc.* **1995**, *9*, 85–91.
10. Knight, L. B., Jr; Steadman, J. *J. Chem. Phys.* **1982**, *77*, 1750–1756.
11. Knight, L. B., Jr; Steadman, J. *J. Chem. Phys.* **1983**, *78*, 6415–6421.
12. Chambers, A.; Fitch, R. K.; Halliday, B. S. *Basic Vacuum Technology*; Institute of Physics Publishing: Bristol, **1989**.
13. Harris, N. S. *Modern Vacuum Practice*; McGraw-Hill: London, **1989**.
14. Hucknall, D. J. *Vacuum Technology and Applications.* Butterworth-Heinemann: Oxford, **1991**.

3

Equipment: ancillaries

The construction of a matrix-isolation cold cell has been fully discussed in Chapter 2. Matrix experiments will normally also require one or more pieces of ancillary equipment. These include

- a preparative vacuum line
- a means of generating reactive species
- one or more spectrometers.

This chapter deals with each of these topics.

1. Preparative vacuum lines

In order to make up gas mixtures for matrix deposition and to carry out other sample manipulations, it is necessary to have a preparative vacuum line. Although this can be mounted on the same trolley as the cold cell, it is better if it is independent of the matrix-isolation system. It can even be in a separate room, provided it is reasonably close to the matrix-isolation laboratory.

A preparative vacuum line need not be elaborate. In the author's laboratory the simple glass line shown in Fig. 3.1 has proved sufficiently versatile over many years. It is constructed out of Pyrex tubing of 14–15 mm diameter and standard vacuum taps and connectors. The following sections cover the features that are needed in a preparative vacuum line for matrix-isolation studies, while more detailed information on the construction and use of vacuum systems can be found in specialist books on the subject.[1-3]

1.1 Vacuum taps and connectors

The vacuum line shown in Fig. 3.1 is fitted with screwed PTFE vacuum taps with captive O-rings. The glass part is divided by vacuum taps into three sections, which can be isolated separately. Each section has two or three inlets fitted with vacuum taps. The inlets for permanently installed attachments, such as gauge heads, are fitted with standard 14/23 ground-glass cones or, in one case, a socket. The glass Pirani gauge head is mounted on the socket inlet with Apiezon W40 wax. The inlets with cones provide mounting points for the other gauges and the flexible line connected to the gas cylinder, all of which

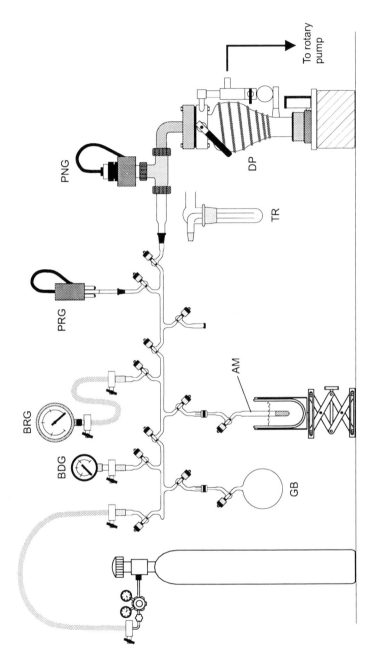

Fig. 3.1 A preparative vacuum line. **AM**, ampoule containing a sample; **BDG**, Bourdon gauge; **BRG**, barometer gauge; **DP**, diffusion pump; **GB**, gas bulb; **PNG**, Penning gauge head; **PRG**, Pirani gauge head; **TR**, trap which can be cooled with liquid nitrogen (optional).

utilize NW10 flanges. Aluminium adapters with an NW10 flange at one end (cf. Fig. 2.5) and a tapered socket to fit 14/23 cones at the other are easily made and are used to mate the various flanged components to the vacuum line. The adapters are sealed onto the glass cones with Apiezon W40 wax.

There are three inlets pointing downwards for the attachment of gas bulbs, sample ampoules and other bits and pieces. These are fitted with high-vacuum screw connectors, which use a PTFE bush to effect the seal.

The older type of greased taps and sockets can be used in place of PTFE taps and screw connectors, but suffer from some disadvantages:

(a) grease quickly becomes contaminated with volatile materials and must be cleaned off and renewed periodically;

(b) high-vacuum grease becomes very stiff when the laboratory temperature is low, making the taps hard to turn; and it runs if it gets too warm, causing leaks to develop;

(c) it is difficult to prevent a coating of grease spreading over the inside of the vacuum line and over the outside of sample ampoules and gas bulbs; so working with the line tends in time to become messy.

Despite these disadvantages many research workers will already have access to lines with greased taps and connections, or may prefer them on grounds of cost or availability. Nor do the advantages lie entirely with screwed taps and connectors. It is more difficult to see whether a screw tap is open or closed than the older type of stopcock, and screw connectors work only with the correct diameter of glass tube with a rather narrow tolerance. If a greased line is to be used, best results will be obtained if the taps are greased following the procedure given in Protocol 1.

Protocol 1.
Cleaning and greasing vacuum taps on a glass line

The following procedure is appropriate for ground-glass vacuum taps con-sisting of a spigot and a barrel and can be carried out without dismantling the vacuum line, provided that extensive internal cleaning of the line is not re-quired. If the line is a new one make sure the correct spigots and barrels are together; they are not always interchangeable. Non-interchangeable taps usually have their parts numbered.

Equipment
• A hair dryer or gentle heat gun

Materials
• High-vacuum grease such as Apiezon L (silicone grease is too mobile)
• Tissues
• Dichloromethane (or petroleum ether)   **harmful (flammable)**

Protocol 1. *Continued*

1. All taps to be greased, both spigots and barrels, should be clean. If the line has been in use for some time and the taps are to be regreased, clean the taps thoroughly with a solvent such as dichloromethane or petroleum ether, using tissues to wipe away any obstinate grease. Ensure adequate ventilation if dichloromethane is used. The ground glass surfaces of the taps should be completely grease free. Clean each tap in turn and replace the ungreased spigot back in its barrel, separating the two parts with a sliver of tissue paper to prevent any chance of seizure. This ensures that spigots and barrels, which are often not interchangeable, are kept together.

2. Once all the taps are clean, carry out the following procedure for each one in turn.

3. Withdraw the spigot from its barrel and remove any tissue placed there to keep it from seizing up. Warm the spigot gently with a heat gun.

4. Using an index finger, spread a small amount of grease over the ground-glass surface of the warmed spigot. The grease should be spread evenly over the whole ground glass surface, but no grease should enter the hole. The amount of grease needed is best judged after some practice, but it should be only enough to coat the spigot thinly. The spigot will appear transparent and only faintly coloured by the grease. Do not apply grease to the tap barrel.

5. Take the spigot in one hand and the heat gun in the other. Hold the spigot near the tap barrel and gently warm both with the heat gun. With the hole in the spigot aligned with the holes in the barrel, insert the spigot neatly and without rotation. Press it gently home to spread the warm grease over the ground glass inner surface of the barrel. Do not rotate the spigot. The whole assembly should appear transparent with an even layer of nearly colourless grease over the ground-glass surfaces. If insufficient grease has been used, it is best to clean the tap and start again.

6. Leave the tap to cool.

1.2 Vacuum pumps and gauges
1.2.1 Pumps

It is generally acknowledged that the preparative vacuum line does not need to attain such a high vacuum as is needed in the cold cell. The line shown in Fig. 3.1 has a diffusion pump similar to those installed on the matrix systems—nominally pumping $135 \, \text{l s}^{-1}$ of air—but a smaller pump will do. High grade gases used for matrices commonly contain impurities in the 10 ppm range. This is equivalent to a partial pressure of impurity of 10^{-3} mbar in a total gas pressure of 100 mbar. The residual pressure in the vacuum line should not add appreciably to the impurity levels in the matrix gases; so the vacuum line should ideally be capable of pressures of 10^{-4} mbar or lower.

A glass trap that can be cooled with liquid nitrogen, shown as an option in Fig. 3.1, can be fitted if desired. It should not be needed, however. All condensable samples should routinely be recondensed into their original containers before these are removed from the vacuum line. There is thus no need to allow significant quantities of condensable materials to escape towards the pumps.

1.2.2 Gauges

Two types of gauge are required: at least one to measure high vacuum and at least one more to measure pressures of gas in the 1–1000 mbar range.

i. High vacuum gauges

High vacuum gauges should be chosen to suit the attainable vacuum. Sensitive Pirani gauges operate in the range 10^{-1}–10^{-4} mbar, and are thus just about suitable for this application. The vacuum line shown in Fig. 3.1 can reach 10^{-7} mbar, so an additional Penning gauge is fitted. The two gauges together cover the range 10^{-1}–10^{-7} mbar, which is more than adequate for the preparative line.

ii. Gauges for measuring gas pressures

Gauges to measure gas pressures are of four main types: Bourdon gauges and barometer gauges, both of which are dial gauges working mechanically; electronic pressure transducers, and conventional fluid-filled manometers.

Bourdon gauges

The simplest and cheapest type of dial gauge is the Bourdon gauge, which has a coiled capillary tube connected to the vacuum system and attached to a pointer. Increasing pressure tends to uncoil the capillary and thus rotate the pointer against its scale. Bourdon gauges are sensitive to changes in atmospheric pressure and should be adjusted to read zero when completely evacuated. This may be necessary more than once a day. Bourdon gauges are available in a variety of pressure ranges, but those with a scale 0–1000 mbar are the most commonly useful.

Barometer gauges

Barometer gauges are independent of atmospheric pressure, and work on the same principle as the familiar aneroid barometer. Once set to zero at high vacuum, they seldom need recalibration. They are also generally more accurate than Bourdon gauges—a model with a quoted accuracy of ±5% of full scale would be typical. A barometer gauge has a sealed corrugated container of gas on the inside, connected to a sensitive pointer. The whole mechanism is enclosed in a vacuum tight chamber, which is opened to the vacuum system. The gas to be measured therefore fills the entire gauge head—a considerable volume. Barometer gauges are heavier and bulkier than Bourdon gauges, as well as being more costly. Like Bourdon gauges, barometer gauges are available in various pressure ranges. There are even expensive types that have a logarithmic scale, capable of reasonable percentage accuracy from 1 to 1000 mbar

Electronic pressure transducers

Modern electronic pressure transducers are much more accurate, but also much more expensive than barometer or Bourdon gauges. They comprise a head unit and an electronic control unit. The control unit gives a digital readout of the pressure. Some models have quoted accuracies of ±0.03% of the reading, while ±0.15% is fairly standard. Models are available in which the head units are resistant to corrosive gases. Gauge heads can be obtained in a variety of ranges from 0–1 mbar up to 0–1000 mbar. For most matrix-isolation work, it would be possible to cope with a single head, reading 0–1000 mbar, but it is preferable to have an additional, more sensitive head. More than one head unit can be run from a single controller. If the most accurate manometry possible is essential, then gauges of this type are the best choice.

Manometers

It is possible to substitute manometers of the conventional fluid type for Bourdon or barometer gauges. Mercury manometers can be used to measure gas pressures up to 1000 mbar, but then need to be nearly one metre tall. Without resort to a cathetometer or other optical magnification, a manometer can be read with an accuracy of about ±0.25-0.5 mm (approximately 0.5 mbar); so the percentage accuracy of a mercury manometer is not very great at pressures below about 5 mbar. The vapour pressure of mercury at room temperature is about 10^{-3} mbar, thus low levels of mercury can be expected to be present in matrices when mercury manometers are used to measure host gas pressures. It is best to avoid even low-level mercury contamination, especially in photochemical experiments where mercury photosensitization could conceivably compete with direct photolysis.

More sensitive manometers, using an oil of low vapour pressure, such as dibutyl phthalate or silicone oil, instead of mercury, are useful for measuring pressures in the range 0.1–2 mbar. At these low pressures they are more accurate than typical barometer gauges, while causing minimal contamination of the matrix gas mixtures. At room temperature, silicone diffusion-pump oils, for example, have vapour pressures below 10^{-5} mbar and relative densities (specific gravities) of about 1.07. A manometer filled with this type of oil is over twelve times more sensitive than if filled with mercury (relative density 13.5).

A simple design for a manometer is shown in Fig. 3.2. This can be filled with oil, for pressure measurements in the range 0.1–2 mbar, or mercury for the range 1–20 mbar. Once the manometer is filled with oil or mercury and attached to the vacuum line, both taps are opened and the entire manometer evacuated down to 10^{-3} mbar or better. One tap is then closed and the other is

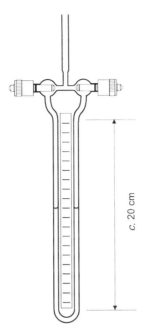

c. 20 cm

Fig. 3.2 A simple glass manometer.

left open. A pressure of gas in the vacuum line will then register as a difference in level between the liquid columns in each side of the manometer, and this can be measured with a millimetre scale or a home-made scale calibrated in mbar.

Choice of gauges

In general it is a good idea to have one gauge for gas pressures of 1–25 mbar or thereabouts and another reading up to 1000 mbar. If an expensive gauge is fitted, careful thought should be given before exposing it to potentially corrosive vapours. The author's group discovered that SO_3 can ruin a barometer gauge after only a few exposures. Bourdon gauges are more robust than barometer gauges and much cheaper to replace. If considerations of cost are not too restrictive, an electronic gauge with two heads of different sensitivities would be ideal. It would be a good idea to keep a Bourdon gauge or manometer for occasional use with doubtful materials, even if more expensive gauges are fitted for normal work.

The vacuum line of Fig. 3.1 has a barometer gauge for the range 1–25 mbar and a Bourdon gauge for 1–1000 mbar; this is a reasonable compromise. The Bourdon gauge is light enough to be fitted directly to the vacuum line with just one clamp around the aluminium adapter with which it is attached. The barometer gauge, being much heavier, is supported independently and connected to the vacuum line via a stainless-steel flexible pipeline. Even if

Bourdon and barometer gauges are installed on the preparative vacuum line for routine use, a manometer of the type shown in Fig. 3.2 is a handy accessory for occasional use. It can be employed for more accurate measurement of low pressures, to calibrate the other gauges, or to avoid contaminating the more expensive dial gauges with corrosive vapours.

1.3 Connections to gas cylinders

Various gases are needed for matrix-isolation studies. Many of these are obtained in standard metal cylinders in a range of sizes, and these have to be connected to the vacuum line. A regulator should always be used, preferably of a type designed for high purity gases, and the regulator should be fitted with a flow-control valve on the outlet. Care should always be taken not to expose a glass line to pressures greater than atmospheric. The vacuum line shown in Fig. 3.1 has a flexible pipeline attached at one end for the purpose of connecting gas cylinders. If gas mixtures are to be made up requiring more than one cylinder, it is convenient to construct some sort of subsidiary manifold line, which can be connected to the glass line via a flexible pipeline. Each cylinder feeding the manifold should have its own regulator and flow control valve. The potential for mistakes is increased with gas manifolds, but in certain circumstances and with care they greatly speed up the preparation of gas mixtures.

1.4 Metal vacuum lines

Some people prefer metal vacuum lines to glass ones. It is straightforward to construct one from stainless-steel tubing and compression fittings. Stainless-steel lines have the disadvantage that visual inspection for internal contamination is not possible, but to compensate they can be heated quite strongly to rid them of adsorbed materials. In laboratories where stainless-steel lines are used, it is routine to leave them under vacuum overnight with a heating tape wound round all or just unused parts of the system.

Vacuum lines made from copper tubing cannot be recommended. Copper is prone to corrosion and reacts with many materials that might be used as matrix guests, often catalysing the decomposition of moderately unstable compounds.

1.5 Spray-on lines for matrix cold cells

The need for a small vacuum line for matrix deposition, which is mounted directly on the matrix-isolation unit, has been dealt with in Chapter 2, while the use of such a spray-on line for depositing matrices is explained in Chapter 4. Glass or metal lines are suitable for this purpose. Figure 4.5 shows a glass vacuum line which provides the minimum facilities needed for a spray-on line. It is a much simplified version of the preparative line illustrated in Fig. 3.1. It has a Pirani gauge for high vacuum monitoring, a Bourdon gauge reading 0–1000 mbar, and an inlet port for attachment of a gas bulb. One end of the vacuum line is connected to the pumping system via a flexible metallic pipeline, and the other to the cold cell via a needle valve and a second flexible line. Needle valves are made to a variety of specifications, and the more expensive types afford finer control of gas flows. Even the cheaper needle valves will be found suitable for matrix deposition in most cases.

2. Generating reactive species

Not all matrix-isolation experiments involve the generation of highly reactive species. For example, molecular complexes, such as those between water or hydrogen halides and a range of organic molecules, can be studied by making up gas mixtures containing the reactants in varying proportions, depositing these as matrices and examining the matrix IR spectra. Nonetheless, the matrix-isolation technique was originally developed for the study of reactive species, and this is probably still its main application.

Many methods of generating highly reactive species have been utilized in matrix studies. These include UV and vacuum-UV irradiation, pyrolysis, X-irradiation, electron and proton bombardment, sputtering, sonication, microwave discharge, and chemical reaction. In a general book on practical matrix isolation, it is impossible to cover all these techniques in detail, some of which are highly specialized. The most widely utilized techniques, and the ones with which the author is most familiar, are photolysis and pyrolysis. Most of the examples to be found in later chapters make use of one or other of these.

When a matrix-isolation system is being set up, it will usually be the case that the intended methods of generating reactive species will have been decided in advance, and the necessary equipment will already be available. If items such as photolysis sources or microwave units are to be purchased specifically for a new matrix laboratory, they will add significantly to the overall budget requirements.

2.1 Photolysis sources
2.1.1 Medium and high pressure arc lamps
All sorts of light sources have been used in matrix experiments, from tungsten-filament bulbs to lasers. The most versatile sources are probably high pressure mercury or mercury–xenon arcs with input powers in the range 200–1000 W, or medium pressure mercury arcs of 100–150 W. These have useful output powers in several wavelength bands spaced over the UV and visible regions of the spectrum. Argon-plasma arc lamps, which have a greater output intensity in the range 210–260 nm than conventional Hg or Hg/Xe arcs, are also used for matrix photolysis, and seem to offer advantages at the shorter wavelengths.

A high pressure arc lamp should be used in conjunction with a convex quartz lens to collimate the beam, at least roughly, a movable shutter, and a water filter of about 5–10 cm pathlength to protect the matrix from unwanted

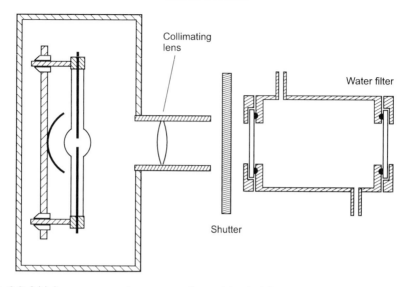

Fig. 3.3 A high pressure arc-lamp source for matrix photolyses.

IR radiation. The shutter need be nothing more complicated than a metal plate, or even a sheet of aluminium foil, which can be moved by hand in and out of the light beam. The water filter should be fitted with quartz or fused silica windows, and these can be seated on O-rings and held in place by screwed cover plates. It should be possible to pass a continuous stream of water through the filter, to prevent it getting hot. Alternatively, the water filter could be constructed with a secondary cooling jacket. A very simple water filter can be made from a 5–10 cm length of glass tubing of about 30 mm diameter, to which appropriately sized silica windows are glued with an epoxy resin. With this design, however, the windows cannot be removed for cleaning. High pressure lamps must be enclosed in a strong housing, to contain all fragments in the event of an explosion. A typical arrangement of a high pressure arc lamp for matrix photolysis is shown in Fig. 3.3.

i. RF interference caused by arc lamps

When a high pressure arc lamp is switched on, a high-voltage spark is generated momentarily to initiate the discharge. At this point a significant level of RF radiation is created, which can interfere with electronic equipment in the vicinity. Computers, spectrometers, and temperature controllers can all be affected. The usual result is just a temporary malfunction of the electronic equipment, which can be corrected by switching it off and then back on again. On occasions, however, it has been known for damage to occur which requires replacement of components.

Therefore, take care when activating a high-pressure arc lamp. It is often not possible to switch off electronic equipment each time an arc lamp is switched on, but, with the lamp on a trolley, it should be possible to wheel it to a point in the laboratory away from sensitive equipment, then wheel it back to the matrix system, once the arc has been struck and is operating in a stable manner.

2.1.2 Low pressure arc lamps

The cheapest sources of UV light are the low pressure mercury arcs that are available in various forms and sizes. The unfiltered, so-called germicidal, lamps produce essentially monochromatic light at 254 nm. The smallest size, consisting of two 4 W tubes in a light aluminium holder, including power supply, is adequate for matrix photolyses. The source of light is spatially extended, so the beam cannot be collimated, but the lamps can be placed fairly close to the external windows of the shroud and reasonably short photolysis times can be attained. If preferred, the light can be passed through a water filter to reduce any tendency for the lamps to warm the matrix. Similar filtered lamps with nominal wavelengths of 312 and 365 nm are also available. These low pressure Hg arcs lack the versatility of high pressure arcs with extended wavelength ranges, but the purchase price and running costs are very much lower, and it is surprising how often 254 nm light is adequate for matrix photolysis.

2.1.3 Matrix photolysis

For maximum convenience, photolysis sources can be mounted on trolleys with castors, so that they can be wheeled up to the cold cell when needed. Provision for adjusting the height of the photolysis beam should be incorporated.

Photolyses with selected wavelength ranges are achieved by use of filters or monochromators. The latter need to be of the high radiance type, with adjustable slits to enable various bandwidths to be selected. Filters are of three types:

(a) glass cut-off filters, which can be either high pass (i.e. allowing wavelengths above a certain value through) or low pass;

(b) interference filters, which allow through only a fairly narrow range of wavelengths and are therefore bandpass filters;

(c) solution filters—often composed of more than one solution—which can be of the high pass, low pass, or bandpass type. The solutions are contained in vessels similar to the water filter shown in Fig. 3.3. Occasionally a gas such as chlorine forms part of a solution filter assembly, and this would need to be contained in a gas cell.

Details of filters can be found in standard works on practical photochemistry.[4,5]

2.2 Pyrolysis tubes

It is possible to hook up the standard commercial type of pyrolysis tube and furnace to a cold cell, but the whole assembly becomes rather bulky. An alternative is to construct a more compact pyrolysis tube. Provided care is taken to protect the hot parts from being accidentally touched, there seems no reason why a simple length of quartz tubing wrapped around with heater wire could not be attached to an inlet port on the shroud, and this would provide the simplest route into pyrolysis experiments with matrix trapping of the products. It is essential that the tube can point directly at the cold window, otherwise pyrolysis products will contact other surfaces before reaching the cold window, with every chance of secondary reaction. The products may even condense there and not reach the cold window at all. With a direct line of sight, the hot tube will inevitably radiate heat towards the cold window, resulting in a temperature rise, but this will not prevent matrix deposition unless the heat radiated becomes excessive.

A variety of designs of pyrolysis tubes and their uses in matrix-isolation studies have been discussed by Moskovits and Ozin.[6]

2.2.1 A water-jacketed pyrolysis tube

In the author's laboratory, a specially made pyrolysis unit, incorporating water cooling, has been constructed from copper and brass. The design of this is shown in Fig. 3.4. The unit is capable of producing temperatures up to about 1000 K, but this maximum could easily be increased.

The outer casing is constructed in three parts: the main body with a water cooled jacket, a middle section with a flange carrying low-current electrical feedthroughs, and an end plate carrying the inlet tube and the electrical terminals for the heater element. The whole assembly is held together by four

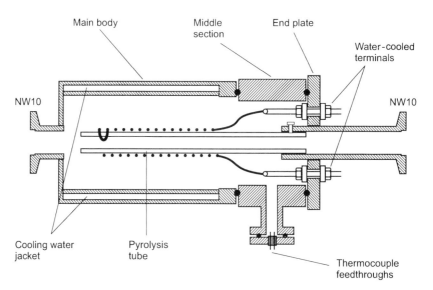

Fig. 3.4 A water-cooled pyrolysis tube for matrix-isolation experiments. A thermocouple (not shown) can be located on the pyrolysis tube with fireproof cement and connected to the outside via electrical feedthroughs.

long screws passing through the end plate, through holes in the casing of the middle section and into threaded holes in the main body. Single O-ring seals ensure vacuum tightness.

The pyrolysis tube, which is made from quartz or a refractory material (e.g. pyrophyllite), has an outside diameter of about 10 mm and fits snugly into the copper inlet tube, which has an NW10 flange at the other end. A single screw, tightened very gently, keeps the pyrolysis tube in place. Because the whole heating unit is located within the vacuum system, fairly thick tungsten or nichrome wire is used as the heating element. Thinner wires have been tried but they quickly break when heated under vacuum. The heating wire is wrapped around the pyrolysis tube as a loop and kept in place with a little fireproof cement. It needs a hefty low-voltage high-current supply, feeding through substantial electrical cables and terminals. The electrical terminals for the heating element pass through the end plate via insulated compression fittings. Each one is a U-tube of 6 mm copper, carrying a flow of cooling water. There are four insulated compression fittings in all: two for each terminal. In some laboratories, pyrolysis tubes are occasionally packed loosely with quartz wool for part of their length to increase the heated surface area.

2.2.2 Measurement of pyrolysis temperatures

A type K thermocouple can be located on the pyrolysis tube by means of a small amount of fireproof cement. It must not touch the heater wire. The thermocouple connections to the outside are taken via the feedthroughs located on a flange on the middle section. Feedthroughs of various types are available, but the easiest to use are small glassy discs with one or more wires passing through, which can be glued into recessed holes in a metal plate with epoxy resin.

An alternative and easier way of assessing the temperature of the hot tube is to calibrate the power supply in separate experiments. A thermocouple is mounted on a blank flange and inserted into the tube so that it contacts the tube wall about midway along the heated section. Vacuum is applied to the system, and the power supply (usually a variable transformer) can then be calibrated, yielding a series of settings needed to achieve given temperatures. This method obviously does not monitor the temperature during the pyrolysis experiments themselves and assumes consistency in the temperatures produced at each setting of the power supply. It is adequate for most purposes, however, and avoids fiddly wiring of permanently installed thermocouples.

The complete pyrolysis unit fits on to an NW10 flange of an inlet port of a vacuum shroud. A matrix host gas, such as argon, can be passed through the hot tube along with the sample to be pyrolysed, or it can be admitted to the shroud independently via a second inlet port.

2.2.3 Pulsed pyrolysis

A recent innovation has been the development of pulsed pyrolysis. This technique was first described by Chen's group in 1992[7] and has since been

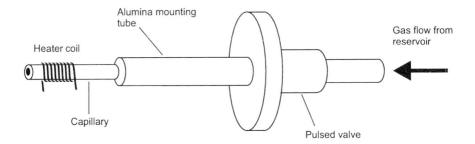

Fig. 3.5 Nozzle design for pulsed pyrolysis. The capillary tube has an i.d. of c. 1 mm and is made of stabilized ZrO_2, SiC or Al_2O_3; it is fixed into an alumina mounting tube with alumina cement. A length of c. 10 mm of the capillary is heated to temperatures up to 2000 K by means of a coil of tungsten or molybdenum wire.

adapted for matrix-isolation studies by Maier, Reisenauer, and co-workers.[8,9] Figure 3.5 shows a design of a nozzle for pulsed pyrolysis, consisting of a heated capillary (c. 1 mm i.d.) a few centimetres long and cemented into a wider alumina tube (aluminium oxide). A length of about 1 cm of the capillary is heated by a tungsten or molybdenum coil to temperatures of up to 2000 K. A mixture of sample and host gas is admitted from a 2 litre reservoir, containing 800–1000 mbar of the gas mixture, in pulses of about 2 msec duration via a solenoid valve. The resulting pulses of pyrolysate are then deposited as a matrix. Pulsed valves can be obtained from specialist suppliers, but the author understands that fuel-injection valves from car engines can be adapted to this purpose, with little or no modification (Mercedes-Benz valves are recommended).

With pulsed pyrolysis, the reaction conditions are quite different from those of more conventional vacuum pyrolysis: the momentary pressure of gas is higher, contact times are briefer, the tube temperatures higher, and, on exiting the heated capillary, the gas mixture expands as a supersonic jet and is thus rapidly cooled. These differences between pulsed and conventional pyrolysis can lead to interesting differences in chemical selectivity.

2.2.4 Reactive tubes

The thermal generation of reactive species prior to matrix deposition is not confined to simple thermolysis reactions. It is possible to pack tubes with a variety of reagents, and allow suitable precursors to pass through them and undergo gas-phase reactions. The reactive products are finally trapped in a matrix. This approach seems to offer a wide scope for generating novel matrix-isolated species.

One example of the application of reactive tubes is the generation of the silicon dihalides, SiF_2, $SiCl_2$ and $SiBr_2$, by passing the corresponding tetrahalides over elemental silicon at 1150–1450 K.[10,11] Matrix IR spectra of the dihalides were obtained using this technique. Phosphoryl chloride (POCl$_3$) has been dechlorinated by reaction with silver at about 1100 K, to give the reactive species POCl, which was then isolated in Ar matrices.[12] Many other examples can be found in the literature.

2.3 Knudsen cells

Knudsen cells are containers made of metal or a refractory material, having an orifice through which a vapour stream can pass when the cell is heated. They find various uses in matrix studies, including

- generation of beams of metal atoms
- carrying out high-temperature reactions, the products from which can be condensed in a matrix
- simple deposition of guest materials of low volatility.

Figure 3.6 shows a Knudsen cell and furnace assembly, which is a simplified version of a design given by Andrews and Pimentel for generating beams of lithium atoms.[13] The cell consists of a stainless-steel body, about 20–25 mm long and 18 mm outside diameter, and a threaded stainless steel plug. The body is drilled out and threaded to accept the plug. The body and plug can be screwed tightly together and sealed with a copper gasket. The plug has a central orifice of 1–2 mm diameter to allow the effusion of vapour generated

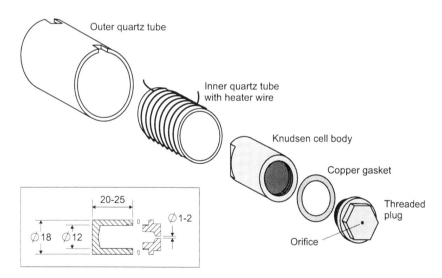

Fig. 3.6 Exploded view of a Knudsen cell and furnace. The cell body and plug are made from stainless steel and, when assembled, fit inside the inner quartz tube. The back of the cell body is drilled with a small hole (not visible in the diagram) to accommodate a thermocouple. The inset shows a section through the cell body, gasket, and plug; dimensions are in mm.

inside the cell. To aid tightening of the plug in the cell body, it should have a square or hexagonal head. For the same reason, the cell body should have a pair of flats machined on the rear, so it can be gripped firmly in a vice. The back of the cell body is drilled with a small hole to take a thermocouple.

The cell is heated by a small furnace consisting of an inner quartz tube with a coil of nichrome wire around it and an outer quartz tube acting as a thermal and electrical insulator. The outer tube is notched at each end to accommodate the ends of the heating wire, which helps to locate the inner tube. In the original design the outer tube was platinum coated to enhance heat retention, but the furnace works adequately up to about 1100 K without this refinement. Heating is achieved by connecting the ends of the heater wire to a standard variable AC supply.

The whole assembly can be incorporated into a stainless-steel shroud such as that shown in Fig. 2.9, by mounting it on a detachable side-plate (Fig. 3.7).The side-plate is conveniently constructed from brass and copper. If desired, the front of the cell can be protected by a radiation shield, made from a small sheet of copper and provided with a small hole, which can be aligned with the orifice in the cell. The shield minimizes radiation from the hot cell and furnace reaching the cold window. The set-up can further be fitted with a simple shutter, moved from outside the shroud by a magnet or by a metal rod passing through a linear or rotary O-ring seal.

The design shown in Figs 3.6 and 3.7 will prove adequate for many applications and is simple to construct. More demanding experiments, especially those requiring very high temperatures (in excess of 1300 K), will need the

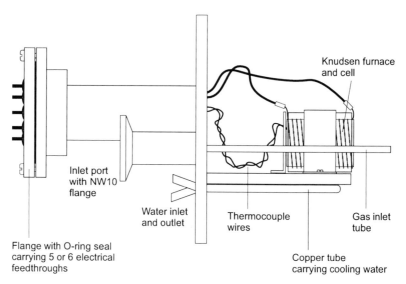

Fig. 3.7 A Knudsen cell and furnace (cf. Fig. 3.6) mounted on a detachable side-plate for fitting to a vacuum shroud such as that shown in Fig. 2.9.

addition of extra radiation shielding and a shutter. Anyone considering constructing a Knudsen cell similar to that shown in Fig. 3.6 should consult the original paper.[13] Also, Moskovits and Ozin have discussed a wide range of applications of Knudsen cells in matrix-isolation studies, and give designs utilizing different materials and methods of heating.[6]

2.4 Microwave discharges

Small microwave sources, which were developed for medical applications, can be used to excite gaseous mixtures and thereby generate reactive species. Commercial microwave sources have small tunable cavities which fit comfortably around Pyrex or quartz tubes of about 15 mm diameter. The conditions in the resulting plasmas are exceptionally drastic, so microwave discharges tend to be used to generate atoms, which are then allowed to react with other species in regions away from the discharge. For more details of the use of microwaves in matrix studies, the reader is referred to recent papers in which sulfur atoms[14,15] and Zn, Cd, and Hg atoms[16,17] were generated by microwave discharge in a tube also incorporating a sample heater.

2.4.1 Argon discharge tubes

One of the most efficient ways of generating cations in matrices makes use of argon passing through a microwave discharge, a technique first developed by Andrews and Prochaska.[18,19] Figure 3.8 shows a vacuum shroud designed for this purpose. Argon is allowed to flow through a quartz tube passing through the microwave cavity, where it is excited. From there it continues through a nozzle of 1–3 mm diameter and is eventually incorporated in the matrix on the cold window as part of the host material. The precursor species is deposited

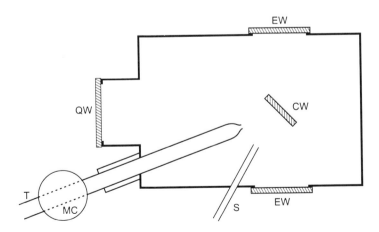

Fig. 3.8 A vacuum shroud for matrix isolation fitted with an argon discharge tube. **CW,** cold window; **EW,** external windows for spectrometer beam; **MC,** microwave cavity; **QW,** outer quartz window for photolysis; **S,** spray-on tube; **T,** quartz discharge tube with nozzle.

simultaneously through a second spray-on tube, either neat or mixed with more of the matrix host gas. In the process, a proportion of the precursor becomes ionized by loss of an electron and the resulting cations are trapped in the matrix. In most cases, an electron acceptor, such as CCl_4, is included with the precursor, and this generally increases the concentration of matrix isolated cations that can be achieved.

With the design of shroud shown in Fig. 3.8, it is possible to conduct an entire experiment without moving the cold cell from the sample compartment of the spectrometer. The cold window can be rotated to the angle shown for matrix deposition and simultaneous argon discharge, or rotated to a position parallel with the external windows (EW) for spectroscopic analysis; while photolysis of the matrix-isolated cations with, for example, a mercury arc can be carried out through the third outer window (QW). It is important not to expose the photomultiplier of a UV–visible spectrometer to high intensity light; so the beam apertures in the sample compartment should be shuttered when the Ar discharge is activated and during photolysis.

There is a long-standing and lively dispute about the mechanism by which ionization of the precursor is brought about in this apparatus. The first possibility is that the argon discharge acts as a vacuum-UV lamp. From this standpoint, the excited Ar atoms emit resonance radiation and the high-energy photons are absorbed by the precursor, resulting in ionization. The vacuum-UV light travels down the quartz tube from the microwave cavity and passes through the nozzle into the shroud. The whole arrangement is therefore a windowless lamp, in which the source of emission also eventually plays the part of matrix host. The second possibility is that excited Ar atoms survive long enough to reach the shroud and undergo gas-phase collisions with the precursor. The resulting energy transfer causes ionization of the precursor. From the practical point of view, it does not matter which of these two ionization mechanisms operates, which is perhaps why there seems to have been no resolution of the issue so far.

3. Spectrometers for matrix isolation

Very many spectroscopic techniques have been applied to the study of low temperature matrices. The most widely used have been IR absorption, UV–visible absorption, and ESR spectroscopy. Others include various types of emission spectroscopy, EXAFS, Mössbauer spectroscopy, magnetic circular dichroism, Raman spectroscopy, and NMR. Some of these techniques are highly specialized and it would be impossible in a general book of this sort to cover them all in detail. A few comments about the most widely used techniques might be helpful, however.

The spectrometers used for matrix-isolation studies do not always need to be highly specified or especially expensive. Because of the size of the matrix cold cell and its attendant trolley system, it is best to use spectrometers in which the optical and control units are in separate cases, otherwise spectrometer controls can be inconveniently placed or even inaccessible once the cell is positioned in the sample compartment. Spectrometers which consist of an optical unit run from a separate computer are ideal. A large sample compartment and high signal-to-noise ratio are certainly advantageous, but good matrix work has been done on quite ordinary spectrometers.

With a spectrometer in which the beam comes to a focus, consideration should be given to the position of the cold window in relation to the focal point. If the beam is focused in the plane of the deposited matrix, only a portion of the matrix will be sampled, and it may be difficult to ensure that the same portion of the matrix is returned to the beam after removal of the cold cell for photolysis or other operations. It is sometimes best to place the cold window away from the focus of the beam and thus sample the whole matrix, or at least a larger portion of it.

A typical matrix is optically no worse than an oil mull, and therefore presents few problems for a standard grating or Fourier transform IR spectrometer. Thin matrices for UV–visible absorption spectrometry can be of a very high optical standard, though scattering of UV radiation can become a problem with less perfect matrices. It is some advantage therefore to have a UV–visible spectrometer with an end-on photomultiplier close to the sample compartment (to minimize losses by scattering).

Individual research workers will be able to decide for themselves if more advanced specifications, such as high resolution or extended wavelength ranges, are needed for particular applications. In many cases, especially when a matrix-isolation system has only recently been set up, it will be a question of making do with what is already available.

There is also the question of whether an IR spectrometer should be purged with nitrogen or purified air, to remove water vapour and CO_2. Since it is impossible to close the lid of the sample compartment of a spectrometer in which a matrix-isolation cell is located, purging can be effected only with the aid of flexible wide-bore tubes which can be attached to the external windows of the vacuum shroud and through which the IR beam passes. This is an inconvenient arrangement, and the purge is broken every time the matrix cell is removed from the spectrometer, for example for photolysis. Purging is therefore recommended for only the most critical experiments.

ESR spectrometers have various sizes of magnets. Some are quite small. When considering the use of an ESR spectrometer, make sure that the cold cell will fit in the available space. Some models have structural bracing above and to the rear of the cavity which cannot be removed but which frustrates attempts to design a simple cold cell to suit. Ingenious custom-built shrouds, with sample rods offset from centre can solve this problem. Specialist matrix ESR workers buy spectrometers with special air gaps between the magnet pole pieces and design shrouds with the ESR cavity incorporated.[20,21]

It is assumed that once the reader has understood the general requirements of setting up a matrix-isolation system, the extra features needed for a particular form of spectroscopy will be readily accommodated. It is usually easiest to design cold cells that are to be used with one form of spectroscopy only.

References

1. Chambers, A.; Fitch, R. K.; Halliday, B. S. *Basic Vacuum Technology*; Institute of Physics Publishing: Bristol, **1989**.
2. Harris, N. S. *Modern Vacuum Practice*; McGraw-Hill: London, **1989**.
3. Hucknall, D. J. *Vacuum Technology and Applications*. Butterworth: Oxford, **1991**.
4. Rabek, J. F. *Experimental Methods in Photochemistry and Photophysics*, Parts 1 and 2. Wiley: Chichester, **1982**.
5. Scaiano, J. C., ed. *Handbook of Organic Photochemistry, Vols I and II*; CRC Press: Cleveland, **1989**.
6. Moskovits, M.; Ozin, G. A., eds. *Cryochemistry*; Wiley: New York, **1976**, Ch. 2.
7. Clauberg, H.; Minsek, D. W.; Chen, P. *J. Am. Chem. Soc.* **1992**, *114*, 99–107.
8. Maier, G.; Preiss, T.; Reisenauer, H. P. *Chem. Ber.* **1994**, *127*, 779–782.
9. Maier, G.; Pacl, H.; Reisenauer, H. P.; Meudt, A.; Janoschek, R. *J. Am. Chem. Soc.* **1995**, *117*, 12712–12720.
10. Hastie, J. W.; Hauge, R. H.; Margrave, J. L. *J. Am. Chem. Soc.* **1969**, *91*, 2536–2538.
11. Maass, G.; Hauge, R. H.; Margrave, J. L. *Z. Anorg. Allg. Chem.* **1972**, *392*, 295–302.
12. Binnewies, M.; Lakenbrink, M.; Schnöckel, H. *Z. Anorg. Allg. Chem.* **1983**, *497*, 7–12.
13. Andrews, W. L. S.; Pimentel, G. C. *J. Chem. Phys.*, **1966**, *44*, 2361–2369.
14. Mielke, Z.; Brabson, G. D.; Andrews, L. *J. Phys. Chem.* **1991**, *95*, 75–79.
15. Brabson, G. D.; Mielke, Z.; Andrews, L. *J. Phys. Chem.* **1991**, *95*, 79–86.
16. Greene, T. M.; Brown, W.; Andrews, L.; Downs, A. J.; Chertihin, G. V.; Runeberg, N.; Pyykkö, P. *J. Phys. Chem.* **1995**, *99*, 7925–7934.
17. Greene, T. M.; Andrews, L.; Downs, A. J. *J. Am. Chem. Soc.* **1995**, *117*, 8180–8187.
18. Prochaska, F. T.; Andrews, L. *J. Chem. Phys.* **1977**, *67*, 1091-1098.
19. Andrews, L. In *Chemistry and Physics of Matrix-Isolated Species*; Andrews, L.; Moskovits, M. ed.; North-Holland: Amsterdam, **1989**, Ch. 2, pp. 15–46.
20. Knight, L. B., Jr; Steadman, J. *J. Chem. Phys.* **1982**, *77*, 1750–1756.
21. Knight, L. B., Jr; Steadman, J. *J. Chem. Phys.* **1983**, *78*, 6415–6421.

4

Basic procedures for matrix preparation

This chapter deals with procedures for sample preparation and the deposition of low-temperature matrices. It is assumed that equipment similar to that described in Chapters 2 and 3 is available. Some additional minor pieces of equipment are described in the following sections.

1. Host gases

1.1 Choice of host gases

Low-temperature matrices consist of a solidified host gas containing a small proportion of a guest material. The choice of host gas will depend on the exact nature of the matrix experiment. Ideally it should condense on the cold window to form a clear, glassy film with good spectroscopic properties. There are two broad classes of matrix experiments:

- those in which the host is chemically inert and serves only to trap the guest species; and
- those in which the host can react with species generated within the matrix.

In most experiments an inert host gas is desired, such as argon or nitrogen, but occasionally a reactive host, such as carbon monoxide, is needed. The noble gases and nitrogen have the advantage of being transparent throughout the UV–visible and mid-IR regions of the spectrum. Most other hosts, such as CO, CH_4, SF_6 and halocarbons, have strong IR absorptions.

Table 4.1 lists some of the matrix hosts that have been employed at one time or another. Some examples of the use of reactive matrices are given later in the book. In most matrix experiments, however, an inert, transparent host is to be preferred, so that the noble gases and N_2 are the most commonly employed host gases. Considerations of cost and convenience further limit the choice to Ar and N_2, unless special experimental needs dictate otherwise. It should be noted that Ne requires a 6 K cryostat for satisfactory matrix deposition, and that Kr and Xe are more expensive than Ar, but can be warmed to higher temperatures, which is occasionally useful. As a rough guide, the temperature (in K) at which a matrix host is to be deposited on a cold window should be

Table 4.1. Thermal properties of some common matrix hosts

Host	T_d (K)[a]	m.p. (K)	b.p. (K)
Ne	10	24.5	27.1
Ar	35	83.9	87.4
Kr	50	116.6	120.8
Xe	65	161.2	166
N_2	30	63.3	77.4
CH_4	45	90.7	109.2
CF_4	—	123	144
CO	35	68.1	81.7
CO_2	63	216.6[b]	194.6[c]
NO	—	109.6	121.4
N_2O	—	182.4	184.7
O_2	26	54.8	90.2
SO_2	—	197.6	263.1
SF_6	—	222.7	209.4[c]
Cl_2	—	172.2	238.6
C_2H_2	—	192.4	189.2[c]
C_2H_4	—	104.0	169.4
C_2F_6	—	179	194

[a] T_d is the temperature at which diffusion of trapped guests first becomes appreciable; T_d data from ref. 2.
[b] At 5.2 bar.
[c] Sublimes.

no more than half the melting point. A brief discussion of the properties of matrix hosts and their suitability for various types of matrix experiment is given in Chapter 1, while more extensive treatments can be found in earlier publications.[1-4]

1.2 Purity of host gases

Host gases should be of the highest purity available; suppliers tend to call these research grade gases. Table 4.2 gives typical analyses of research grade Ar and N_2. As can be seen, reactive impurities in research grade Ar and N_2, such as CO, H_2, and O_2, can be expected to be present individually at no more than 2 ppm, and the total impurity level should be no more than 10 p.p.m.. Other research grade gases, such as Ne, Kr, Xe, CO, and O_2, are usually supplied with slightly higher impurity levels than Ar and N_2. It is worth reflecting that 100 mbar of research grade Ar or N_2 admitted into a vacuum system with a background pressure of 10^{-3} mbar will gain up to 10 ppm of impurities from the vacuum line. This exceeds the impurity level in the gas as supplied, so vacuum lines with minimum pressures of 10^{-4} mbar or less are recommended for all aspects of matrix work.

Table 4.2. Typical analyses of research grade gases (impurities in p.p.m.)[a]

	Ar	N$_2$	O$_2$
Purity (%)	>99.9995	>99.9995	>99.996
Ar	—	<5.0	<15
CO$_2$	<0.5	<0.5	<0.5
CO	<1.0	<1.0	<1.0
H$_2$	<2.0	<2.0	—
CH$_4$	<0.5	<0.5	<0.5
N$_2$	<3.0	—	<15
N$_2$O	<0.1	<0.1	<0.1
O$_2$	<1.0	<1.0	—
H$_2$O	<0.5	<0.5	<0.5
Total hydrocarbon	<0.5	<0.5	<1.0

[a] Data from Air Products Ltd., quoted with permission.

It is not unknown for water levels in research grade gases to exceed the specified maximum and some matrix workers, in order to ensure freedom from water, routinely pass host gases through copper coils immersed in liquid nitrogen before use. If such a procedure is adopted, the gas must be warmed to room temperature again before manometric measurements are made on the vacuum line.

1.3 Gas cylinders

Research grade gases are relatively expensive and are seldom supplied in normal, full size cylinders. Nevertheless, gases that are to be used as routine matrix hosts are best purchased in cylinders containing at least 100–250 litres, which will last for many experiments. Gases to be used only occasionally can be bought in smaller sizes, right down to glass flasks containing only 1 litre, or even less, at atmospheric pressure.

A cylinder of gas should be connected to the vacuum line via a pressure regulator, preferably of the type which allows purging of the air space within the regulator, and a flow-control valve (see Fig. 4.1). It is important not to contaminate a high purity gas by inadvertently mixing it with small volumes of air.

2. Manipulating gases and volatile materials

2.1 The matrix ratio

The mole ratio of the host gas to guest is called the *matrix ratio*. The normal range for matrix ratios (host:guest) is 100:1–10000:1, although matrix ratios well outside these limits are sometimes employed. It is not always possible to estimate matrix ratios, though it is preferable to do so. Mixing the matrix host gas with the guest material can be achieved in either of two ways, depending on the volatility of the guest.

(a) If the guest is reasonably volatile—having a room-temperature vapour pressure of about 1 mbar or more—the host and guest can be mixed in the gas phase on a preparative vacuum line such as that shown in Fig. 3.1, transferred to the matrix-isolation unit in a gas bulb, and deposited as a gas mixture. The matrix ratio can be determined by simple manometry on the preparative vacuum line.

(b) If the guest is less volatile, it is usually necessary to evaporate it from a side-arm attached to the vacuum shroud of the matrix cell and deposit the host gas separately. With this deposition method, it is much more difficult to estimate matrix ratios.

The next few sections describe in detail the various techniques and pieces of equipment needed for the preparation of matrix-gas mixtures and the deposition of matrices.

2.2 Vacuum lines for gas handling

In order to deposit a low-temperature matrix, it is necessary to mix both the host gas and the guest material in the vapour phase. There are numerous ways in which the gases may be manipulated, but the most versatile involves the use of a separate preparative vacuum line (cf. Fig. 3.1) and a spray-on line mounted on the matrix-isolation unit (cf. Figs 2.13–2.15). In the descriptions of procedures which follow, it will be assumed that some such arrangement is available. None the less, readers will readily see how the functions of preparative and spray-on lines could be combined into a single vacuum line mounted on the matrix-isolation unit. Though not so convenient as separate lines, such a combined set-up can offer worthwhile economies of laboratory space as well as capital and maintenance costs.

2.3 Gas bulbs and sample ampoules

2.3.1 Gas bulbs

Matrix host gases should always be deposited from a reservoir vessel, such as a glass gas bulb (see Fig. 4.1), and not directly from the cylinder. This allows measurement of the quantity of gas admitted into the vacuum shroud of the cold cell and minimizes the risk of allowing large volumes of gas into the shroud by mistake. Glass bulbs can be made in a variety of sizes, but for general use 1 litre bulbs are best. Smaller sizes are useful for expensive materials, such as isotopic gases or guest materials available only in very small amounts. Larger bulbs are useful when the host gas has a relatively low vapour pressure at room temperature. For instance, 3-methylpentane forms nice glassy matrices, but it is a liquid at room temperature with a vapour pressure of about 150 mbar. Bulbs larger than 2 or 3 litres tend to be awkward when mounted on the matrix-isolation system, however, and it may be more convenient to refill a smaller bulb during a matrix experiment if such large volumes are required.

Gas bulbs should be equipped with high quality vacuum taps and the appropriate connectors for attaching them to the preparative vacuum line and spray-on line. As a safeguard against the risk of injury from glass fragments in the event of an implosion, glass gas bulbs should be criss-crossed over their entire surface with adhesive tape or otherwise protected. If light-sensitive compounds are to be used, gas bulbs should be covered in aluminium foil or other opaque wrapping.

One of the simplest and most frequently needed operations in matrix-isolation research is the filling of a gas bulb with a known pressure of gas. Protocol 1 describes this procedure.

Note that it is unwise to allow the pressure in a glass vacuum line to exceed atmospheric, even by a small amount. Apart from the danger of bursting at high pressures, even small over-pressures can put strain on waxed seals, resulting in leaks.

Protocol 1
Filling a gas bulb

The following is the procedure for filling a bulb to a given pressure with gas from a cylinder.

Caution! Wear eye protection at all times when working on a glass vacuum line.

Equipment
- A preparative vacuum line such as that shown in Figs 3.1 and 4.1
- A 1 litre gas bulb

Materials
- A cylinder of the required gas

1. Use a preparative vacuum line like the one shown in Fig. 4.1, and refer to this figure when following the protocol. (The sample ampoule and Dewar vessel are not needed for this procedure.)
2. Check that the vacuum line is working properly and is evacuated to an acceptable pressure (10^{-3} mbar or better).
3. Connect a 1 litre gas bulb to a convenient port on the vacuum line (GB connected to T8 in Fig. 4.1). If it is intended to fill the bulb with gas to more than 500 mbar, support it with clamps or a lab-jack. At lower fill pressures, the atmosphere will keep the bulb supported.

Protocol 1 *Continued*

4. Connect a cylinder of the required gas, such as research grade N_2, to another port (T1), via a pressure regulator (GR), flow-control valve (FCV), and flexible transfer line (TL).

5. Arrange that the gas pressure regulator is filled with the gas up to the flow-control valve; this may need to be done before connecting the gas-transfer line. The flow-control valve should be sufficiently gas-tight to support a vacuum on one side and more than one atmosphere of gas on the other, without appreciable leakage.

6. First evacuate the air from the gas bulb, the space between T8 and T9, and the gas transfer-line, as follows.

  (a) Isolate the gauges from the line during air pumping, by closing T2, T4, and T6.

  (b) Isolate the diffusion pump from the pumping system and connect the vacuum line only to the rotary backing–roughing pump.

  (c) Open T8, then T9, then T1, to evacuate the gas bulb and the gas-transfer line as far as the flow-control valve.

  (d) When the normal backing pressure has been restored, return the diffusion pump to the circuit.

  (e) Open T2 and T6, to reconnect the Bourdon and Pirani gauges to the line, and continue pumping with the diffusion pump until a good vacuum (10^{-3} mbar or better) is achieved.

  (f) Test for leaks by closing tap T7 (leaving the Pirani gauge head on the enclosed portion of the vacuum line) and monitor the pressure increase with the Pirani gauge. A slow increase in pressure is normal, but a rapid increase denotes a leak, which should be identified and cured before proceeding further.

  (g) If no serious leak is detected, reopen T7 and proceed.

7. Next fill the gas bulb to the correct pressure, as follows.

  (a) Close T3, to isolate that portion of the line containing the gas bulb, gas-transfer line, and Bourdon gauge (0–1000 mbar).

  (b) Check that T1, T2 and T8 are open, while T9 remains closed.
NB It is vital that T1 and T2 are open at this stage, otherwise the pressure rise in the vacuum line will not be registered and safe pressures may be exceeded.

  (c) Ensure that the flow-control valve is shut and the main valve on the gas cylinder open, then adjust the control knob on the regulator to give a small positive pressure reading on the outlet pressure gauge.

  (d) Gently open the flow-control valve until a gradual increase in pressure is registered by the Bourdon gauge on the vacuum line.

(e) Once a gentle flow of gas has been established, open T9.

(f) By carefully controlling the flow-control valve, allow the pressure in the vacuum line to rise to the required level. Do not exceed 1000 mbar.

(g) Quickly close both the flow-control valve and T9—as nearly simultaneously as possible. This becomes easy with practice.

8. The gas bulb is now ready for use in a matrix experiment. Close T8 and remove it from the preparative line.

9. Leave the vacuum line ready for future use:

    (a) close the main valve on the gas cylinder and check that the regulator is fully off and the flow-control valve closed;

    (b) check that the taps on ports, such as T8 and T10, are all closed;

    (c) isolate the diffusion pump from the circuit and evacuate the excess gas from the entire vacuum line, including all gauges, using only the rotary backing pump;

    (d) return the diffusion pump to the circuit;

    (e) check that there are no serious leaks and leave the vacuum line under pumping.

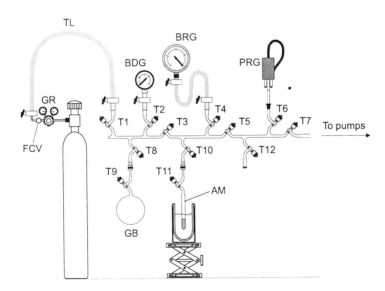

Fig. 4.1 A preparative vacuum line set up for matrix-gas preparation. **AM**, ampoule containing the volatile guest material; **BDG**, Bourdon gauge; **BRG**, barometer gauge; **FCV**, flow-control valve; **GB**, gas bulb; **GR**, gas-pressure regulator; **PRG**, Pirani gauge head; **TL**, flexible metal gas-transfer line.

2.3.2 Bulbs with break-seals

Very expensive gases, such as those with isotopic substitution (e.g. ^{13}CO or $^{15}N_2$) are usually supplied in glass bulbs of 100-cm^3 to 1 litre capacity. These come with a *break-seal*, and must be fitted with a vacuum tap and a means of breaking the seal before they can be used. Figure 4.2 shows how this is done. The neck of the bulb is extended with a short length of glass tubing of appropriate diameter, terminating with a vacuum tap. The glass extension tube has a hooked side-tube, in which an iron, steel or brass ball is placed. It is important that the side-tube is wide enough round the bend to allow the ball to be rolled into the neck of the gas bulb.

Before use, the air space between the vacuum tap and the break-seal is evacuated. The vacuum tap is then closed and the ball used to break the internal glass seal. If an iron or steel ball is used, this can often be done without removing the bulb from the vacuum line, by lifting the ball with a small magnet and dropping it onto the seal. The seal sometimes survives one impact, however, and it may be necessary to remove the bulb from the vacuum line and give it one sharp, downwards shake.

Once the seal is broken, the bulb should always be kept with the vacuum tap uppermost. This prevents the small fragments of glass that come from the broken seal coming into contact with the vacuum tap. The lower sealing surfaces of PTFE valves are very easily damaged by broken glass, resulting in leaks.

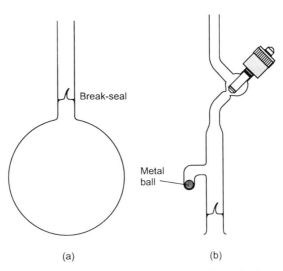

(a)   (b)

Fig. 4.2 A glass bulb with a break-seal: (a) as supplied, (b) the neck after modification for use.

2.3.3 Sample ampoules

Reasonably volatile guest materials, that is, those with a room-temperature vapour pressure of about 1 mbar or more, are conveniently stored in Pyrex ampoules fitted with high quality vacuum taps and connectors appropriate to the vacuum lines on which they are to be used. A useful size can be made out of pyrex tubing of 14–15 mm o.d., which should be long enough to stand upright in a standard Dewar vessel with the tap well above the neck—say 25–30 cm overall (see Fig. 4.1). Even fairly unstable guest materials, such as diazo compounds, azides and peroxides, can then be stored for weeks at liquid-nitrogen temperature, provided the discipline of keeping the storage Dewars topped up is maintained. In this way, guest materials can be kept readily available for matrix experiments.

Many samples can be stored in ampoules, either at room temperature or in a freezer, without deteriorating excessively. If freezer storage is contemplated, make sure that the vacuum tap fitted to the sample ampoule will not develop a leak when cooled. Some taps with metal cores have this disadvantage, and should be avoided for ampoules which are to be refrigerated.

2.4 Making up matrix-gas mixtures

One of the most common procedures in matrix-isolation experiments is the making up of gas mixtures. These may be mixtures of two or more gases, such as CO or O_2 in Ar, or mixtures of the vapours of liquid or solid guest materials with host gases. Protocol 3 describes the procedure for preparing a mixture of N_2 and acetone (N_2:acetone $= 100:1$). This involves degassing the acetone (described in Protocol 2) and mixing its vapour with the host gas. The procedure is quite typical and may be applied to the preparation of any mixture of the vapour of a liquid or solid guest with a gaseous host. When a matrix-isolation system is first set up, it is recommended that the system be tested as soon as possible by preparing a straightforward gas mixture, such as acetone in N_2 or Ar, and depositing the mixture as a matrix (see Protocol 7).

2.4.1 Degassing a sample

Before a volatile liquid sample can be used in preparing a matrix-gas mixture, it should be degassed by freeze–pump–thaw cycles. This will remove any dissolved or entrapped air and also any gaseous products resulting from de-composition of the sample, for example N_2 from azides or diazo compounds. It is most important that degassing is carried out thoroughly, otherwise residual air may form the predominant component in the vapour above the sample. Protocol 2 describes the procedure for degassing a liquid. Volatile solid samples should be degassed similarly, but may require fewer cycles than liquids.

Protocol 2
Degassing a liquid sample

The following is the procedure for degassing a liquid sample, usually the guest material for a matrix experiment. The procedure may be applied equally well to the degassing of a solid, except that no bubbling will occur. In the case of compounds with a risk of explosion, such as azides, diazo compounds, and peroxides, some extra precautions are necessary, and these are noted in the protocol.

Caution!   Wear eye protection at all times when working on a glass vacuum line.

Equipment
- A preparative vacuum line such as that shown in Figs 3.1 and 4.1
- A sample ampoule
- A wide-mouth Dewar vessel

Materials
- The liquid sample
- Liquid nitrogen   **very low temperature, danger of severe frostbite**

1. Place 1–5 ml of the liquid to be degassed into the sample ampoule with the aid of a teat pipette. Depending on the type of vacuum tap fitted to the ampoule, this may be best achieved by first withdrawing the tap piston. Try to minimize accumulation of droplets of the liquid in the vicinity of the tap, by inserting the tip of the pipette as far into the ampoule as possible. Replace the tap piston, if it has been removed, and close the tap.

    NB If this procedure is adopted with a potentially explosive material, such as a diazo compound or azide, put no more than 1 or 2 ml of the sample into the ampoule, place a safety screen in front of the ampoule after it is connected to the vacuum line, use heavy duty gloves when manipulating the line, and wear goggles or a safety mask. With low molecular weight azides and diazo compounds, where the risk of explosion is greatest, wear ear protectors as well.

2. Check that the vacuum line is working correctly and evacuated to an acceptable pressure (10^{-3} mbar or better).

3. Connect the sample ampoule to the vacuum line at one of the vacant ports, as shown in Fig. 4.1 (T10). (The gas bulb, gas cylinder, and transfer line are not needed for this procedure.)

4. First evacuate the air from the space between T10 and T11 and from the ampoule, by the following sequence.

    (a) Isolate as much of the line as possible and all the gauges, while the air is being pumped out, by closing T3, T4, and T6.

(b) Isolate the diffusion pump from the circuit and, using only the rotary backing pump, evacuate the air space between T11 and T10, by opening T10.

(c) Cool the sample ampoule with liquid nitrogen in a wide-mouth Dewar vessel. The sample will be cold enough once vigorous boiling of the liquid nitrogen has subsided.

(d) Carefully open T11 and pump out the air in the ampoule with the rotary backing pump, until the normal backing pressure has been restored.

(e) Keeping the ampoule cold, return the diffusion pump to the circuit and open T6, to allow readings to be made with the Pirani gauge.
NB If potentially explosive materials are being used, on no account allow sample vapour to contact Pirani or Penning gauge heads; so make sure that the sample is not opened to the gauge until it is fully cooled.

(f) Continue pumping until a reasonably good vacuum (10^{-3} mbar or better) has been attained.

5. Next allow the sample to warm and release entrained air into the head-space in the sample ampoule.

(a) Close T11. This is very important to prevent vapour from reaching the gauge heads or the pumps.

(b) Remove the Dewar of liquid nitrogen from around the ampoule and warm the sample to room temperature. The warming can be speeded up by immersing the tip of the ampoule in a beaker of cold water and swirling gently. Avoid warming the ampoule with the fingers, since frostbite can result.

(c) Check that T11 is fully closed by monitoring the pressure recorded by the Pirani gauge. There should be no increase.

(d) Allow the sample to reach room temperature. As the sample warms, dissolved air will bubble out of solution. Eventually the sample may itself boil under the low pressure in the ampoule. It is not always possible to distinguish bubbles of air from bubbles caused by boiling of the sample.

6. Finally get rid of the released air.

(a) Immerse the tip of the ampoule in liquid nitrogen again, and cool the sample thoroughly.

(b) Open T11 and note the response of the Pirani gauge to the small amount of air released.

(c) Continue pumping the sample at liquid-nitrogen temperature until a good vacuum (10^{-3} mbar or better) has been attained.

(d) Close T11.

7. Repeat the freeze–pump–thaw cycles until no perceptible dissolved air

Protocol 2   *Continued*

remains. Judge this from the response of the Pirani gauge after the ampoule has been cooled with liquid nitrogen in each cycle and T11 is reopened. The response should be negligible.

8. Close T11 for the last time, and then T10.   The sample is now ready for preparation of a matrix-gas mixture. It may be warmed to room temperature or left at liquid nitrogen temperature for the time being.

9. Return the vacuum line to its normal state, with all parts under vacuum, in readiness for its next job.

2.4.2 Preparing a matrix-gas mixture

As an example of the preparation of a matrix-gas mixture, Protocol 3 describes the procedure for mixing the vapour of a volatile liquid guest (acetone) with a typical host gas (N_2). Acetone is chosen as a conveniently available stable compound with a reasonably high vapour pressure and a series of strong characteristic IR absorptions. It is thus an ideal guest compound for testing a matrix-isolation system. The procedure is based on three principles:

(a) it is best to pump each section of the vacuum line until the last possible moment.

(b) portions of the vacuum line which do not need to be exposed to organic or other vapours should not be exposed to them.

(c) portions of the vacuum line which have been in contact with organic or other vapours should be isolated from the system as soon as they are no longer needed.

The procedure may be applied equally well to the preparation of other gas mixtures.

Protocol 3
Preparing a mixture of acetone and N_2 (N_2:acetone $=$ 100:1)

The following procedure involves the degassing of the guest material (acetone) and the preparation of a mixture of the guest with the host gas, N_2. It may be readily adapted to mixtures of other volatile guests (both solid and liquid) and other host gases. In the case of compounds with a risk of explosion, however, such as azides and diazo compounds, some extra precautions are advisable. These are noted in the protocol.

Caution!   Wear eye protection at all times when working on a glass vacuum line.

4: Basic procedures for matrix preparation

Equipment

- A preparative vacuum line such as that shown in Figs 3.1 and 4.1, including a flexible metal gas-transfer line and a pressure regulator with flow-control valve
- A 1 litre gas bulb
- A sample ampoule
- A wide-mouth Dewar vessel

Materials

- Acetone of analytical grade (1–5 ml)   **flammable**
- Research grade N_2 (in a cylinder)
- Liquid nitrogen   **very low temperature, danger of severe frostbite**

1. Use a preparative vacuum line like the one shown in Fig. 4.1, and refer to this figure when following the protocol.

2. Check that the vacuum line is working normally and is evacuated to an acceptable pressure (10^{-3} mbar or better).

3. Connect a 1-litre gas bulb to a convenient port (GB connected to T8 in Fig. 4.1), and a cylinder of research grade N_2 to another port (T1), via a pressure regulator (GR), flow-control valve (FCV) and flexible transfer line (TL).

4. Arrange that the gas pressure regulator is filled with N_2 up to the flow-control valve; this may need to be done before connecting the gas-transfer line. The flow-control valve should be sufficiently gas-tight to support a vacuum on one side and more than one atmosphere of N_2 on the other, without appreciable leakage.

5. First evacuate the air from the gas bulb, the space between T8 and T9, and the gas transfer line, as follows.

   (a) Isolate the gauges from the line during air pumping, by closing T2, T4, and T6.

   (b) Using only the rotary backing pump (keeping the diffusion pump out of circuit), open T8, then T9, then T1, to evacuate the gas bulb and the gas-transfer line as far as the flow-control valve.

   (c) When the normal backing pressure has been restored, return the diffusion pump to the circuit.

   (d) Open T2, T4 and T6, and continue pumping with the diffusion pump until a good vacuum (10^{-3} mbar or better) is achieved.

   (e) Test for leaks by closing tap T7 (leaving the Pirani gauge head on the enclosed portion of the vacuum line) and monitor the pressure increase with the Pirani gauge. A slow increase in pressure is normal, but a rapid increase denotes a leak, which should be identified and cured before proceeding further.

   (f) If no serious leak is detected, reopen T7 and leave the line under pumping, while the sample ampoule is prepared.

6. Place 1-5 ml of the acetone in the sample ampoule with the aid of a teat pipette.

Protocol 3 *Continued*

NB If this procedure is adopted with a potentially explosive material, such as a diazo compound or azide, put no more than 1 or 2 ml of the sample into the ampoule, place a safety screen in front of the ampoule after it is connected to the vacuum line, use heavy duty gloves when manipulating the line, and wear goggles or a safety mask. With low molecular weight azides and diazo compounds, where the risk of explosion is greatest, wear ear protectors as well. On no account allow vapour from potentially explosive samples to contact Pirani or Penning gauge heads.

7. Close T3, to keep air away from the gas bulb, and connect the ampoule to the vacuum line at a suitable port (T10 in Fig. 4.1).

8. Evacuate the ampoule and degas the sample by the method described in Protocol 2. At the end of the procedure, make sure T11 is closed, and warm the degassed acetone to room temperature.

9. Make sure the rest of the vacuum line is still properly evacuated:

    (a) leave T11 closed, open T3 again, and pump the remainder of the vacuum line, including the gas bulb, all gauges and the transfer line;

    (b) check that no leaks have developed by monitoring the pressure.

10. Next fill the gas bulb to the correct pressure with acetone, as follows.

    (a) Close T1, T2, and T5, to isolate that portion of the line containing the gas bulb, sample ampoule, and barometer gauge. (T3 and T4 should remain open.)

    (b) Carefully open T11 very slightly until a pressure begins to register on the barometer gauge (BRG).

    (c) Allow the pressure to rise to 5.0 mbar, then close T11. It should be possible to control the pressure to 0.5 mbar or less. If the required pressure is accidentally exceeded, condense the acetone back into the ampoule by immersing the tip of the ampoule in liquid nitrogen, then begin to fill the bulb again.

    NB If a compound other than acetone is being used, the maximum attainable vapour pressure may be less than 5 mbar.

    (d) Close T9, leaving the gas bulb containing 5 mbar of acetone.

11. Remove excess acetone from the rest of the vacuum line, in the following sequence:

    (a) Immerse the tip of the sample ampoule in liquid nitrogen, open T11, and condense the acetone in the remainder of the vacuum line back into the ampoule.

    (b) Leave the liquid nitrogen in place until the barometer-gauge reading returns to zero. There is no need to allow appreciable amounts organic vapours to pass through the pumps.

(c) Close T11 and open T5. Traces of residual acetone will now be pumped from the system and will register on the Pirani gauge.

(d) Open T11 and continue pumping until a reasonable vacuum (10^{-3} mbar or better) has been attained.

(e) Close T11 and T10, leaving the sample cold and under vacuum. The sample ampoule can be warmed and removed from the line at this stage, but it is more sensible to leave it in place until the whole procedure has been completed, in case of accidents and the need to repeat the operation.

12. The remainder of the procedure is to make up the gas bulb to the correct pressure with N_2. This is similar to the procedure for filling a gas bulb described in Protocol 1, but with the additional need to ensure efficient mixing of the acetone and N_2. Carry this out using only the portion of the vacuum line beyond T3, as follows.

(a) Close T4. This prevents any residual adsorbed acetone in the barometer gauge and its connecting tube from bleeding into the system.

(b) Open T1 and T2; there should be no more than a slight rise in the Pirani gauge reading.

(c) Continue pumping until a good vacuum (10^{-3} mbar or better) has been attained. At this stage the vacuum line, including the Bourdon gauge and transfer line, are under vacuum, while the gas bulb still contains acetone and the gas regulator is filled with N_2.

(d) Close T3 to isolate the working portion of the line.

(e) Check that T1, T2, and T8 are open, while T9 remains closed. It is very important that both T1 and T2 are open at this stage, so that the pressure rise will be registered on the Bourdon gauge.

(f) Ensure that the flow-control valve is shut and the main valve on the gas cylinder open, then adjust the control knob on the regulator to give a small positive pressure reading on the outlet pressure gauge.

(g) Gently open the flow-control valve until a gradual increase in pressure is registered by the Bourdon gauge on the vacuum line. The aim is to fill the gas bulb with 500 mbar of N_2, with optimum mixing of the acetone and N_2 and minimum effusion of acetone from the gas bulb. (Strictly speaking the final pressure should be 505 mbar, but the gauge will probably not allow an accuracy of better than ±5–10 mbar.)

(h) By carefully controlling the flow-control valve, allow the pressure in the vacuum line to rise to about 800 mbar. Do not exceed 1000 mbar.
NB This is greater than the final pressure required in the gas bulb, but gas at this pressure is confined to a small volume of vacuum line.

(i) Close the flow-control valve, then open T9 and reclose it quickly. This allows N_2 to enter the gas bulb in a rush, ensuring turbulent mixing with

Ian R. Dunkin

Protocol 3 *Continued*

the acetone. The pressure in the line should drop well below 500 mbar. If not, rapidly open and close T9 again.

(j) Carefully admit more N_2 into the line by opening the flow-control valve again, and repeat the rapid opening and closing of T9. On each occasion, leave T9 open only for the shortest time possible. As this process is continued, the pressure in the gas bulb will approach the target of 500 mbar.

(k) Complete the filling process by opening the flow-control valve very slightly and allowing a very small flow of gas into the line. This will be registered as a gradually increasing reading on the Bourdon gauge. Then open T9 and keep it open until the desired pressure of 500 mbar is reached.

(l) Quickly close both the flow-control valve and T9—as nearly simultaneously as possible.
NB If the initial vapour pressure (p) of the guest material was different from 5 mbar and a matrix ratio (host:guest) of 100:1 is desired, the final pressure, once host gas has been admitted, should be $100 \times p$ (or more precisely $101 \times p$).

13. The gas mixture is now ready for use in a matrix experiment. Close T8 and remove the gas bulb from the preparative line.

14. It is good practice to leave the vacuum line ready for future use.
    (a) Check that T11 is closed.
    (b) Warm the sample ampoule to room temperature and remove it to safe storage.
    NB If a sample other than acetone is used which is unstable, leave it cold and transfer the sample ampoule to a storage Dewar containing liquid nitrogen or to a freezer.
    (c) Close the main valve on the gas cylinder and check that the regulator is fully off and the flow-control valve closed.
    (d) Check that the taps on ports, such as T8 and T10, are all closed.
    (e) Isolate the diffusion pump from the circuit and evacuate excess N_2 from the entire vacuum line, including all gauges, using only the rotary backing pump.
    (f) Return the diffusion pump to the circuit.
    (g) Check that there are no serious leaks and leave the vacuum line under pumping. Even relatively innocuous materials such as acetone can take an appreciable time to be desorbed from internal surfaces, but the pressure in the line should return to its normal minimum in a few hours. If this does not occur, use gentle heating of affected parts of the vacuum line, to assist desorption of contaminants.

2.4.3 Organic vapours as host gases

Protocol 3 can be easily adapted to the preparation of gas mixtures consisting of a volatile guest material and a host gas which is the vapour of an organic solvent, such as 3-methylpentane. The procedure is the same except that

(a) the gas cylinder is replaced by a second ampoule containing the host material as a liquid. This can be attached to the vacuum line via T12 in Fig. 4.1, or more logically via T10, with the sample ampoule connected via T12;

(b) the solvent to be used as host must be degassed by freeze–pump–thaw cycles (Protocol 2), as well as the guest.

(c) the final pressure of the mixture will be limited by the room-temperature vapour pressure of the host, 150 mbar in the case of 3-methylpentane, for example. This needs to be determined before the appropriate pressure of guest material can be ascertained;

(d) a gas bulb larger than 1 litre may be desirable in order to ensure an adequate supply of the matrix-gas mixture;

(e) care should be taken to condense host vapour back into its ampoule before re-opening the vacuum line to the pumps.

2.4.4 Mixtures of two or more cylinder gases

It is sometimes necessary to prepare mixtures of gases to be used as more complex hosts, for example small percentages of CO or O_2 in Ar or N_2. Protocol 3 can again be adapted for this purpose. The minor component can be treated as the guest, and the procedure is the same, except that

(a) no degassing procedure is required;

(b) neither component can be condensed, so excess of each will need to be removed through the pumps—backing pump alone first;

(c) at least two gas cylinders will be involved—one for the major component and one for the minor. These can be accommodated in one of the following ways:

   i.   Connect the cylinder containing the minor component to the vacuum line first (as shown in Fig. 4.1), fill the gas bulb to the required pressure, then replace the cylinder with the second one, containing the major component.

   ii.   Connect the second cylinder via a free inlet port on the vacuum line; so that both are installed on the vacuum line together.

   iii.   Build a separate gas manifold, for example out of copper tubing and compression fittings; so that more than one gas cylinder, each with its own pressure regulator, can be connected to the vacuum line via a single port.

These options increase in convenience of use but also in expense (*i-iii*). The last can be recommended if mixtures of cylinder gases are going to be needed frequently. Even so, a gas manifold needs to be used with care if cross contamination of gases is to be avoided.

This procedure can obviously be adapted to the preparation of mixtures of three or more gases, but careful thought needs to be given to the order of mixing in such cases.

2.4.5 Dilute matrix mixtures

The procedure of Protocol 3 is adequate for mixtures with matrix ratios up to about 500, but more dilute mixtures require the addition of an expansion bulb to the standard set-up. The use of expansion bulbs is explained in Sections 2.5 and 2.6.

2.5 Expansion bulbs

When dilute matrices are needed—those with matrix ratios greater than about 500—it is necessary to introduce very low pressures of guest material into the gas bulb, before topping up with host gas. A typical barometer gauge reading 0–25 mbar can give adequate pressure readings down to about 1 mbar (though with diminishing accuracy at the lower end), while an oil manometer, such as that shown in Fig. 3.2, might be used down to about 0.1 mbar. Pressures below these require the use of expansion bulbs. Even when slightly higher pressures—in the range 1–5 mbar—are needed, they may often be obtained with greater accuracy by means of expansion bulbs.

No attempt should ever be made to use Pirani, Penning, or other high vacuum gauges to measure low pressures of volatile guest materials. This is because

(a) they are usually calibrated for air, and have quite different sensitivities to other gases;

(b) they are quickly contaminated by organics, inorganics, and organo-metallics and soon malfunction;

(c) they can detonate potentially explosive compounds.

On the last point, the author has evidence that one serious vacuum-line explosion was caused by his inadvertently allowing a low pressure of diazomethane—admittedly a very hazardous compound—into a Pirani gauge head. The resulting explosion fragmented a portion of the vacuum line, the sample ampoule and the surrounding Dewar vessel. Injury was avoided, because most of the safety precautions recommended in Protocols 2 and 3 had been adopted, namely a safety screen, a face mask, and heavy duty gloves.

Expansion bulbs are fitted between the vacuum line and the gas bulb in which the matrix-gas mixture is to be prepared (see Fig. 4.3). They are simply small bulbs equipped with male and female vacuum connectors. As with all

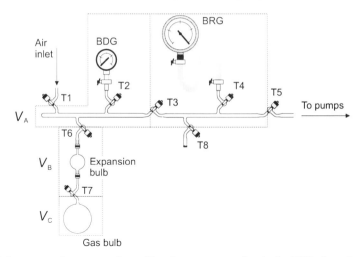

Fig. 4.3 A vacuum line set up for calibrating an expansion bulb. **BDG**, Bourdon gauge; **BRG**, barometer gauge. The volumes used in calibrating the expansion bulb are shown as V_A, V_B, and V_C. The inclusion of the barometer gauge in V_A is optional.

vacuum components, they can be made of glass or metal, as preferred. Expansion bulbs can be made in many different sizes, but the most generally useful are those that are approximately 10 and 100 times smaller than the gas bulb with which they are to be used.

2.5.1 Calibration of expansion bulbs

Before an expansion bulb is used, it should be calibrated, that is, the ratio of its volume to that of the gas bulb determined. There are two main methods of calibrating expansion bulbs, one involving three pressure measurements and the other four pressure measurements. Only the second method is suitable for the smaller sizes of expansion bulbs. The two methods are described in Protocols 4 and 5. There is no need to use expensive research grade gases—or indeed any cylinder gases—for calibration: air will do just as well.

Protocol 4
Calibrating an expansion bulb: method 1

This procedure enables the volume ratio of a gas bulb and an expansion bulb to be determined, and is suitable for fairly large expansion bulbs, i.e. those of 100 cm^3 or more. For smaller expansion bulbs, method 2 is more suitable (see Protocol 5).

Caution! Wear eye protection at all times when working on a glass vacuum line.

117

Ian R. Dunkin

Protocol 4 *Continued*

Equipment

- A preparative vacuum line similar to that of Figs. 3.1 and 4.3
- A gas bulb
- An expansion bulb

1. Refer to Fig. 4.3, which shows a section of the preparative vacuum line. Decide at the outset whether one or more pressure gauges are needed. Unless the gas bulb is very large, a single gauge, reading 0–1000 mbar, will probably suffice. If two gauges are needed, use the section of vacuum line between taps T1 and T5; if only one is needed, use the section between T1 and T3. For the rest of this protocol, it is assumed that only one gauge will be used. Nevertheless, if two gauges are to be used, leave both open to the vacuum line at all times. This is to maintain constant volume. Do not worry if the more sensitive gauge exceeds its maximum reading; it will have been designed to withstand pressures up to atmospheric.

2. Check that the vacuum line is working properly and is evacuated to an acceptable pressure (10^{-3} mbar or better).

3. Connect the gas bulb and expansion bulb as shown in Fig. 4.3, leaving taps T6 and T7 closed.

4. The pressure in the system will rise to near atmospheric at some points during the procedure, so support the bulbs gently to prevent them falling off. This is especially important if greased cones and sockets are used to make the connections rather than screw couplings.

5. Ensure that tap T1 is closed and T2 open, then isolate the diffusion pump from the pumping system, and evacuate the section of vacuum line between T1 and T3 (including the gauge), using the backing pump only.

6. Still using only the backing pump, open T6 then T7 to evacuate the expansion bulb and gas bulb.

7. If no leaks are apparent, return the diffusion pump to the circuit, and continue pumping until a reasonably good vacuum has been has been attained (10^{-3} mbar or better).

8. Now isolate the portion of the vacuum line to be used, by closing T7, T6, and T3, preferably in that order.

9. Admit air into the section of line enclosed by T1, T6, and T3 (this includes the Bourdon gauge), by carefully opening T1.

10. Allow the pressure to rise to about 7–800 mbar, then close T1.

11. Record the pressure: P_1.

12. Leaving T1, T3 and T7 closed, open T6 to allow the air to expand into the expansion bulb. Record the pressure: P_2.

13. Open T7 and record the final pressure: P_3.

118

14. The three pressure readings, P_1, P_2 and P_3, are made with the air confined to volumes V_A, $V_A + V_B$ and $V_A + V_B + V_C$, respectively. The volume ratio required, that of the gas bulb to the expansion bulb, is given by V_C/V_B, and from Boyle's law,

$$\frac{V_C}{V_B} = \frac{P_1(P_2 - P_3)}{P_3(P_1 - P_2)}.$$

---

i. Example of calibration by method 1

For calibration by method 1, assume typical values of V_A, V_B and V_C equal to 250, 100 and 1000 cm^3, respectively. Given these volumes, reasonable values for the experimentally measured pressures are

$$P_1 = 800, \quad P_2 = 570, \quad P_3 = 150 \text{ mbar.}$$

A high quality gauge should be accurate to ± 1–2% of full scale; so these pressure readings should be accurate to ± 10-20 mbar. From the equation given in Protocol 4, the ratio of the volumes of the gas bulb and expansion bulb is

$$\begin{aligned}
\frac{V_C}{V_B} &= \frac{P_1(P_2 - P_3)}{P_3(P_1 - P_2)} \\
&= \frac{800 \times 420}{150 \times 230} \\
&= 9.7
\end{aligned}$$

which is near enough to the correct ratio of 10. Given the likely errors, the volume ratio determined in this way should be accurate to about $\pm 10\%$, which is adequate for most purposes. If the ratio V_B/V_A is small, however, i.e. if the expansion bulb is small, the percentage error in the difference, $P_1 - P_2$, will become relatively large. In these circumstances calibration of the expansion bulb by method 2 is to be preferred (Protocol 5).

Protocol 5
Calibrating an expansion bulb: method 2

This procedure enables the volume ratio of a gas bulb and an expansion bulb to be determined, and is suitable even for small expansion bulbs.

Caution! Wear eye protection at all times when working on a glass vacuum line.

Equipment
- A preparative vacuum line similar to that of Figs 3.1 and 4.3
- A gas bulb
- An expansion bulb

Protocol 5 *Continued*

1. Refer to Fig. 4.3, which shows a section of the preparative vacuum line. Decide at the outset whether one or more pressure gauges are needed. For small expansion bulbs, two gauges will usually be required for sufficient accuracy, one reading 0–1000 mbar and the other 0–25 mbar. If two gauges are needed, use the section of vacuum line between taps T1 and T5. If only one is needed, use the section between T1 and T3. For the rest of this protocol, it is assumed that two gauges will be used. Leave both gauges open to the vacuum line at all times. This is to maintain constant volume. Do not worry if the more sensitive gauge exceeds its maximum reading; it will have been designed to withstand pressures up to atmospheric.

2. Check that the vacuum line is working correctly and is evacuated to an acceptable pressure (10^{-3} mbar or better).

3. Connect the gas bulb and expansion bulb as shown in the Fig. 4.3, leaving taps T6 and T7 closed. The pressure in the system will rise to near atmospheric at some points during the procedure, so support the bulbs gently to prevent them falling off. This is especially important if greased cones and sockets are used to make the connections rather than screw couplings.

4. Ensure that taps T1 and T8 are closed and T2, T3 and T4 open, then isolate the diffusion pump from the pumping system and evacuate the section of vacuum line between T1 and T5 (including the gauges), using the backing pump only.

5. Still using only the backing pump, open T6 then T7, to evacuate the expansion bulb and gas bulb.

6. Wait until the normal backing pressure is restored, then, if no leaks are apparent, continue pumping with the diffusion pump, until a good vacuum has been attained (10^{-3} mbar or better).

7. Close T7 and T5, preferably in that order.

   NB The expansion bulb is still open to the vacuum line, but the gas bulb is not.

8. Admit air into the section of line enclosed by T1, T7, T8, and T5 (this includes the expansion bulb and both gauges), by carefully opening T1.

9. Allow the pressure to rise to about 7–800 mbar, then close T1.

10. Record the pressure: P_1.

11. Close T6 and, using first the backing pump and then the diffusion pump, evacuate the section of line (including the gauges) between T1 and T5. This leaves air in the expansion bulb only.

12. Close T5, then open T6, allowing the air in the expansion bulb to expand into the vacuum line. Measure the pressure: P_2.

13. Leaving T6 open, admit more air into the system via T1, allowing the pressure to rise to 7–800 mbar again. Close T1 and record the pressure: P_3.

14. Open T7 and record the pressure: P_4.

15. From Boyle's law the volume ratio of the expansion bulb and gas bulb, V_C/V_B, is given by

$$\frac{V_C}{V_B} = \frac{P_1 + P_2}{P_2} \times \frac{P_3 - P_4}{P_4}$$

---

ii. Example of calibration by method 2

As an example of calibration by method 2, values of V_A, V_B and V_C are assumed to be 1000, 10, and 1000 cm^3, respectively. Given these volumes, reasonable values for the sequence of measured pressures are:

$$P_1 = 800, \quad P_2 = 8.0, \quad P_3 = 800, \quad P_4 = 400 \text{ mbar.}$$

The low pressure reading (P_2) would be taken from a sensitive gauge such as a barometer gauge with a scale of 0–25 mbar; P_2 would thus be accurate to about ±0.25–0.5 mbar. The other pressures, read from a gauge with a scale 0–1000 mbar, are likely to be accurate to ±10–20 mbar. From the equation given in Protocol 5, the ratio of the volumes of the gas bulb and expansion bulb is

$$\frac{V_C}{V_B} = \frac{P_1 + P_2}{P_2} \times \frac{P_3 - P_4}{P_4}$$
$$= 101 \times 1.00$$
$$= 101$$

which is near enough the true value of 100. By using a sensitive gauge for measurement of the low pressure (P_2), the accuracy of the volume ratio determined in this way can be kept within ±10%.

2.5.2 Accuracy of calibration

Sticklers for accuracy will note that manometer gauges (cf. Fig. 3.2) are generally more accurate than dial type gauges and that oil-filled manometers are particularly sensitive. Ideally each expansion bulb should be calibrated with each gas bulb with which it might be used, and a record kept of the ratios. With screw or push-fit vacuum couplings, a standard degree of insertion should be adopted, to maintain consistency of relative volumes. Finally, it should be noted that all pressure gauges change their internal volumes between zero and full scale. Bourdon gauges probably have the smallest volume change, while with barometer and manometer gauges the effect may be quite significant. In the case of manometers, volume changes can be minimized by constructing them from narrow-bore tubing.

All these sources of calibration error should be considered, but it is futile to pretend that very great accuracy can be achieved in expansion-bulb calibration. The 10% accuracy, which can be achieved fairly easily, will suffice for most purposes.

2.6 Dilute matrix-gas mixtures

With the aid of an expansion bulb, matrix-gas mixtures with matrix ratios (host:guest) of greater than 500:1 may be easily prepared. As an example, Protocol 6 gives the procedure for preparation of a mixture of acetone and N_2 with N_2:acetone = 1000:1. The procedure can be adapted for the preparation of mixtures of other volatile guest materials and other matrix hosts.

Protocol 6
Preparing a mixture of acetone and N_2 (N_2:acetone = 1000:1)

The following procedure involves the degassing of the guest material (acetone) and the preparation of a mixture of the guest with the host gas, N_2. It may be readily adapted to mixtures of other volatile guests (both liquid and solid) and other host gases. In the case of compounds with a risk of explosion, such as azides and diazo compounds, some extra precautions are advisable. These are noted in the protocol.

Caution! Wear eye protection at all times when working on a glass vacuum line.

Equipment

- A preparative vacuum line such as that shown in Figs 3.1 and 4.4, including a flexible metal gas-transfer line and a pressure regulator with flow-control valve
- A 1 litre gas bulb
- A calibrated expansion bulb of approximately 100 cm³ (i.e. 1/10th the volume of the gas bulb)
- A sample ampoule
- A wide-mouth Dewar vessel

Materials

- Acetone of analytical grade (1–5 ml)   **flammable**
- Research grade N_2 (in a cylinder)
- Liquid nitrogen   **very low temperature, danger of severe frostbite**

1. Use a preparative vacuum line like that shown in Fig. 4.4, and refer to this figure when following the protocol.
2. Check that the vacuum line is operating correctly and is evacuated to an acceptable pressure (10^{-3} mbar or better).
3. Connect a 1 litre gas bulb (GB1) to a convenient port via the expansion bulb (EB) and support the bulbs gently by clamping or with a lab-jack.
4. Connect a cylinder of research grade N_2, via a gas pressure regulator (GR), flow-control valve (FCV), and flexible transfer line (TL), as shown in the figure.
5. Arrange that the gas pressure regulator is filled with N_2 up to the flow-

control valve; this may need to be done before connecting the gas-transfer line. The flow-control valve should be sufficiently gas-tight to support a vacuum on one side and more than one atmosphere of N_2 on the other, without appreciable leakage.

6. First evacuate air from the bulbs and the gas-transfer line, as follows.
   (a) Close T2, T4, and T6, to isolate the gauges from the line while air is evacuated.
   (b) Using only the rotary backing pump (keeping the diffusion pump out of circuit), evacuate the gas bulb, expansion bulb, and the gas-transfer line as far as the flow-control valve, by opening T8, T9, and T1.
   (c) Return the diffusion pump to the circuit and open T2, T4, and T6. Continue pumping until a good vacuum (10^{-3} mbar or better) is achieved.
   (d) Test for leaks by closing tap T7 (leaving the Pirani gauge head on the enclosed portion of the vacuum line) and monitor the increase in the Pirani gauge reading. A slow increase in pressure is normal, but a rapid increase denotes a leak, which should be identified and cured before proceeding further.
   (e) If no serious leak is detected, open T7 and leave the line under pumping, while the sample ampoule is prepared.

7. Place 1–5 ml of the acetone in the sample ampoule with the aid of a teat pipette, connect the ampoule to the vacuum line and degas it by the method described in Protocol 2. With T11 closed, warm the degassed acetone to room temperature at the end of the procedure.

   NB If this procedure is adopted with a potentially explosive material, such as a diazo compound or an azide, put no more than 1 or 2 ml of the sample into the ampoule, place a safety screen in front of the ampoule when it is connected to the vacuum line, use heavy duty gloves when manipulating the line, and wear goggles or a safety mask. With low molecular weight azides and diazo compounds, where the risk of explosion is greatest, wear ear protectors as well. On no account allow potentially explosive vapours into contact with Pirani or Penning gauge heads.

8. Ensure that the rest of the vacuum line is still properly evacuated:
   (a) Leave T11 closed (the sample ampoule) and pump the remainder of the vacuum line, including the gas bulb, expansion bulb, all gauges, and the transfer line.
   (b) Check that no leaks have developed by monitoring the pressure.

9. Next fill the expansion bulb to the correct pressure with acetone, as follows.
   (a) Close T1, T2, T9, and T5 (in that order), to isolate that portion of the line containing the expansion bulb, sample ampoule, and barometer gauge. (T3 and T4 should remain open.)
   (b) Carefully open T11 very slightly until a pressure begins to register on the barometer gauge.

Ian R. Dunkin

Protocol 6 *Continued*

    (c) Allow the pressure to rise to 5.5 mbar, then close T11. It should be possible to control the pressure to 0.5 mbar or less. If the pressure is accidentally exceeded, condense the acetone back into the ampoule by immersing the tip of the ampoule in liquid nitrogen, then begin to fill the bulb again. If the volume ratio of gas bulb and expansion bulb differs from 10, adjust the pressure of acetone vapour in the expansion bulb to give a final pressure of 0.5 mbar after expansion.

    NB If a compound other than acetone is being used, the maximum attainable vapour pressure may be less than 5.5 mbar.

    (d) Close T8, leaving the expansion bulb containing 5.5 mbar of acetone.

10. Open T9, allowing the acetone vapour to fill the gas bulb and expansion bulb. With a volume ratio of gas bulb/expansion bulb = 10, this will result in a final pressure of 0.5 mbar of acetone.

11. Remove excess acetone from the vacuum line:

    (a) Close T9 and open T8.

    (b) Immerse the tip of the sample ampoule in liquid nitrogen and condense all the acetone in the remainder of the vacuum line back into the ampoule, by opening T11. Leave the liquid nitrogen in place until the barometer-gauge reading returns to zero. There is no need to allow appreciable amounts organic vapours to pass through the pumps.

    (c) Close T11 and open T5. Traces of residual acetone will now be pumped from the system and will register on the Pirani gauge.

    (d) Open T11 and continue pumping until a reasonable vacuum (10^{-3} mbar or better) has been attained.

    (e) Close T11 and T10, leaving the acetone cold and under vacuum. The sample ampoule can be warmed and removed from the line at this stage, but it is more sensible to leave it in place until the whole procedure has been completed, in case of accidents and the need to repeat the operation.

12. The remainder of the procedure is to make up the gas bulb to the correct final pressure—500 mbar—with N_2. This should be carried out using the portion of the vacuum line beyond T3, following exactly the same procedure as described in instructions 12 and 13 of Protocol 3.

13. Finally leave the vacuum line ready for further use, as described in instruction 14 of Protocol 3

2.6.1 Very dilute matrix-gas mixtures

Very dilute matrix-gas mixtures can be prepared by an extension of Protocol 6. For instance, a mixture of N_2 and acetone with N_2:acetone $= 10^6$:1 can be obtained as follows.

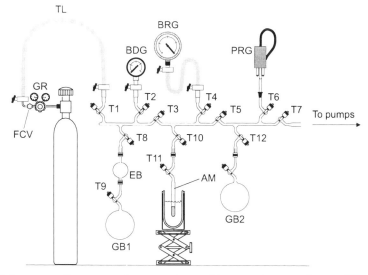

Fig. 4.4 A vacuum line set up for preparing a dilute matrix-gas mixture. **AM**, sample ampoule; **BDG**, Bourdon gauge; **BRG**, barometer gauge; **EB**, expansion bulb; **FCV**, flow-control valve; **GB1**, gas bulb for final mixture; **GB2**, gas bulb for intermediate mixture; **GR**, gas-pressure regulator; **PRG**, Pirani gauge head; **TL**, flexible metal gas-transfer line.

(a) Prepare a mixture with N_2:acetone $= 1000$:1, as described in Protocol 6.

(b) Move the gas bulb containing the mixture to the position of GB2 in Fig. 4.4 and put a fresh gas bulb in the position of GB1. The expansion bulb should remain in place.

(c) Use the procedure of Protocol 6 to expand 5.5 mbar of the gas mixture in GB2 from the expansion bulb into GB1. This will give 0.5 mbar of the first gas mixture in GB1.

(d) Dilute the mixture with pure N_2 to 500 mbar, as described in Protocols 3 and 6. GB1 will now contain the desired mixture with a matrix ratio of 10^6.

2.7 The homogeneity of matrix-gas mixtures

When matrix-gas mixtures are prepared, it is important that they should be homogeneous. In the author's laboratory, and others with which he is familiar, the normal practice is to try to admit host gas into the gas bulb with a certain amount of turbulence, as described in Protocols 3 and 6. This seems to provide well-mixed gases, and no problems with gas inhomogeneity have arisen in the author's experience.

In contrast, some matrix workers believe that homogeneous gas mixtures are guaranteed only when mechanical mixing is employed. This can be achieved by constructing gas bulbs with small magnetic turbines enclosed. The turbines can then be driven by external rotating magnets to ensure complete mixing of the enclosed gases. It seems advisable, however, to wait for evidence that mechanical mixing is actually necessary in a series of experiments before resorting to this expedient on a routine basis.

3. Deposition of matrices

3.1 General deposition methods

There are two ways in which pure gases or gas mixtures can be deposited as low-temperature matrices:

- slow spray-on
- pulsed deposition.

Both make use of a spray-on line connected to the matrix-isolation cold cell (see Chapter 2, Section 7.2.4 and Chapter 3, Section 1.5). Figure 4.5 is a diagram of a glass line of the sort currently in use in the author's laboratory. It uses the same diffusion and rotary pumps as the cold cell itself. A manifold, like that shown in Fig. 2.11, and fitted with valves on two ports, enables the two functions of the vacuum system to be carried out quite separately.

3.1.1 Slow spray-on

In the slow spray-on technique, a gas bulb is filled with a pure gas or a gas mixture, as described in the previous section, and is then mounted on the spray-on line (at T4 in Fig. 4.5). The whole spray-on line (except the gas bulb)

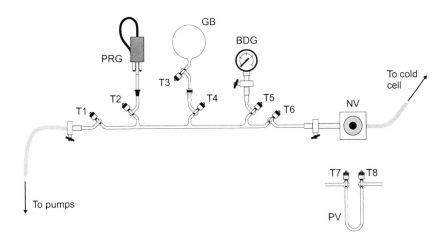

Fig. 4.5 A glass spray-on line. **BDG,** Bourdon gauge; **GB,** gas bulb containing pure host gas or matrix-gas mixture; **NV,** needle valve; **PRG,** Pirani gauge head; **PV,** optional pulsing volume for pulsed matrix deposition.

is evacuated through T1. After a reasonable vacuum has been attained, T1 is closed, and the matrix is deposited on the cold window through the needle valve (NV). Spray-on rates vary widely, but are typically in the range 1–10 mmol h^{-1}. A 1 litre bulb filled to 500 mbar contains 22 mmol of gas; so half the contents would typically be sprayed on in 1–10 hours.

3.1.2 Pulsed deposition

In the pulsed deposition method, the needle valve (NV) is replaced by two normal vacuum taps separated by a *pulsing volume* (Fig. 4.5, PV). Matrix deposition is achieved by opening and closing the first tap (T7) to fill the pulsing volume, then opening and closing the second tap (T8) to allow a pulse of gas into the vacuum chamber of the cold cell. After an interval of a few seconds, or longer, a second pulse is let into the cold cell, and the process repeated until the matrix has been built up to the required thickness. The pulsing volume can be varied, but is usually 10–30 cm^3. The deposition of large numbers of pulses can be tedious and, if the taps are stiff, uncomfortable. If preferred, therefore, pulsing can be automated to a certain extent. Solenoid valves can be used instead of ordinary vacuum taps, and these can be operated electrically, either with manual switches or by a sequence timer. The latter needs to control four timed events, with at least the waiting period between pulses variable:

open T7 – close T7 – open T8 – close T8 – variable delay – repeat sequence.

A means of setting the total number of pulses to be deposited is also needed.

3.1.3 A comparison of slow and pulsed deposition

The slow spray-on technique is by far the most widely used of the two methods. The pulse technique is only applicable where the matrix host and guest are pre-mixed in the gas bulb. If a relatively involatile guest has to be sublimed on to the cold window and the host gas deposited simultaneously, the continuous deposition provided by slow spray-on is essential. The same applies to deposition of reactive species generated externally in the gas phase.

The pulse technique is nevertheless favoured in some laboratories whenever it can be used. This is for two reasons:

(a) Pulsing seems to give more transparent matrices with some host gases. This is thought to be due to a slight annealing of the matrix near the surface as each pulse is deposited, an effect which in some circumstances seems to be beneficial.

(b) Pulsing minimizes the effect of trapping residual gases—principally air and water vapour—which are always present at very low pressures in the vacuum chamber of the cold cell. With slow spray-on, the residual gases are deposited continuously with the host and guest, and so become part of the matrix. With pulsed deposition, the matrix is deposited in a series of quick condensations, with relatively long gaps between. The residual gases are for the most part deposited in thin layers between the individual layers of matrix formed by each pulse. Most of the trapped air and water is therefore kept away from the intended guest species, unless the matrix is warmed sufficiently to allow extensive diffusion.

It is often claimed that, at a shroud pressure of 10^{-6}–10^{-7} mbar, a monolayer of water is deposited on a matrix cold window about every second; so pulsed deposition could significantly improve the degree of isolation of a reactive species.

For the large majority of matrix experiments, however, the only possibility will be the slow spray-on method. In many laboratories pulsing is never used. Whichever method is employed, some workers deposit matrices at the base temperature of the cold cell, commonly 10–14 K, while others prefer to warm the cold window slightly, to 20 K perhaps, deposit at this temperature, and cool the matrix to the base temperature only after deposition is complete. It is worth making some trials to determine which conditions give the highest quality matrices with each matrix-isolation system and in each series of experiments.

Protocol 7 gives the procedure for the slow spray-on of a pre-mixed matrix gas, such as the mixtures of acetone and N_2, which can be prepared following Protocols 3 (N_2:acetone = 100:1) and 6 (N_2:acetone = 1000:1). The best way of monitoring deposition of acetone–N_2 mixtures is by IR spectroscopy, as described in the protocol. It is recommended that a new matrix-isolation system should be tested by depositing a stable mixture of this type at the earliest opportunity.

Protocol 7
Slow deposition of an acetone-N_2 mixture

This is a key part of the matrix-isolation technique: the deposition of a matrix by the slow spray-on method. Acetone–N_2 has been chosen as a convenient and stable mixture for testing purposes, but the procedure may be readily adapted to mixtures of other volatile guests (both solid and liquid) and other host gases.

Caution! Wear eye protection at all times when working with glass vacuum lines.

Equipment

- A complete matrix-isolation system, such as those described in Chapter 2, operating with a closed cycle refrigerator, and fitted with a CsBr, CsI or KBr cold window, KBr external windows, and a spray-on line with an appropriate needle valve.
- A 1 litre gas bulb
- An IR spectrometer capable of accommodating the cold cell in its sample compartment.

Materials

- A gas-phase mixture of acetone and N_2 made up in the gas bulb according the procedures of Protocol 3 or 6 (suggested ratio N_2:acetone = 100–1000:1)

1. It is assumed that the bulb containing the gas mixture for matrix deposition can be transferred to the spray-on line with the cold window already at low temperature, and that the necessary evacuation of air spaces and other gas manipulations can be carried out on this line without affecting the vacuum in the cold cell. With some systems, this may not be possible, and in that case the mixture of acetone and N_2 will have to be made up in the bulb and transferred to the spray-on line before cooling commences.

2. Ensure that the matrix system has no obvious faults:

   (a) Check that the helium pressure in the compressor unit falls within the operating limits specified by the manufacturer. This will typically be in the range 180–220 p.s.i. (12–15 bar). Excessively low pressure will result in the compressor cutting out, and should be rectified by topping up with helium before cooling is attempted.

   (b) Check that the electrical connection between the compressor and head units of the refrigerator is securely plugged in at both ends. If the rotary valve in the head unit is not activated, excessive pressure can build up in the compressor unit when it is switched on, leading to loss of helium through the safety valve and cut-out.

   (c) Check that the cooling water supply, if there is one, is properly connected to the compressor unit.

   (d) Check that the vacuum system is pumping the cold cell and providing a good vacuum (10^{-4} mbar or better).

   (e) Check visually that the external windows and cold window are clean.

   (f) Check that the temperature controller is switched on and giving a reading close to room temperature, and that the selected final temperature is either 0 K or a value below the base temperature of the cold cell.

3. Switch on the refrigerator.

4. In the first few minutes of the refrigerator operating:

   (a) check that the head unit is functioning correctly by feeling the outer case. A regular pulse, caused by the reciprocating pistons inside the unit, should be detectable. If there is no pulse, switch off the refrigerator, check the electrical connection between the compressor and head units, and try again;

   (b) make sure that the temperature of the cold window is falling steadily.

5. The matrix system can now be left to cool to its base temperature. Typically, this will take an hour or two.

6. While the cold window is being cooled, prepare a gas mixture of acetone

Protocol 7 *Continued*

and N_2 as described in Protocols 3 or 6. Any matrix ratio in the range N_2:acetone $= 100:1–1000:1$ will suffice for this test procedure.

7. Transfer the bulb containing the gas mixture to the spray-on line (refer to Fig. 4.5 for this part of the procedure) and set up the line for matrix deposition:

   (a) Attach the gas bulb containing the acetone–N_2 mixture to a vacant port on the spray-on line (T4 in Fig. 4.5) and support it if necessary.

   (b) Ensure that the needle valve (NV) is closed.

   (c) If a single pumping system provides the vacuum for both cold cell and spray-on line, close off the main valve to the cold cell.

   (d) Using the rotary backing pump first, then the diffusion pump, evacuate the spray-on line through T1, including all gauge heads, the volume between T6 and the needle valve, and the air-space between T4 and T3.

   (e) When a good vacuum has been attained (10^{-4} mbar or better), the main vacuum valve to the cold cell can be reopened.

   (f) The cold cell and spray-on line can be left under simultaneous pumping until it is desired to start matrix deposition. Alternatively, the spray-on line can be filled with the matrix-gas mixture at this stage (see below).

8. When the base temperature of the cold window has been attained, record a background IR spectrum of the cold window, as follows.

   (a) Check that the cold window is parallel to the external windows through which the spectrometer beam will pass. Rotate the cold window to this position if necessary.

   (b) Carefully position the cold cell in the sample compartment of the spectrometer. Depending on the type of spectrometer, the optimum position of the cold cell can be judged by maximizing the transmission through the cell at a convenient wavenumber, or by maximizing the interferogram displayed on the screen. It may be possible to deposit the matrix without moving the cold cell from the sample compartment of the spectrometer. If not, ensure that the cold cell can be repositioned reasonably consistently (to within 1–2 mm) each time it is moved.

   (c) Record the background spectrum and check that no gross impurity absorptions are present. With experience, normal background absorptions for a particular cold window will come to be recognized. Abnormally intense absorptions or absorptions in unfamiliar parts of the spectrum usually indicate contamination of the matrix system, which may include built up deposits on the external windows.

   (d) The background spectrum can be subtracted from or used as the reference for all subsequent matrix spectra.

NB In order to make sure that any condensable impurities in the system do not go undetected, or are confused with the matrix sample, it is important to wait until the base temperature has been reached before recording a background spectrum. A background should be recorded for every matrix experiment.

9. Prepare the system for matrix deposition, which should be done without removing the cold cell from the spectrometer sample compartment if possible.

   (a) Rotate the cold window to face the inlet port in the vacuum chamber through which the matrix gas mixture will enter. (Some workers prefer to deposit matrices with the window at an angle (45°, say) to the inlet port.)

   (b) If deposition is to be carried out above the base temperature, adjust the temperature controller to the desired setting, and wait a few minutes to allow the selected temperature to stabilize.

   (c) On the spray-on line (Fig. 4.5), check that there is still a good vacuum with T2, T4, T5, and T6 all open, then close T1, to isolate the line from the main pumping system.

   (d) Close T2 to protect the Pirani gauge head from organic vapours.

   (e) Check that the needle valve is closed, then open T3 to allow the gas mixture into the spray-on line.

   (f) Record the pressure reading on the Bourdon gauge (BDG).

10. Override manually any safety device which would be activated by a pressure rise in the vacuum shroud.

11. Deposit the matrix, using the following sequence.

    (a) Close T3, to isolate temporarily the gas bulb from the line. Deposition will now take place from a relatively small volume, and the pressure drop registered on the Bourdon gauge will be greater for a given amount of gas deposited.

    (b) Carefully open the needle valve (NV) to allow the gas mixture into the cold cell. With experience, the appropriate amount of rotation of the needle valve will be known. Otherwise, err on the safe side, open it gradually in stages, and allow a minute or two between each rotation, to see if any pressure drop is discernible on the Bourdon gauge.

    (c) Once spray-on is underway, monitor the rate of deposition by periodically recording an IR spectrum of the developing matrix. To do this, rotate the cold window to its optimum position in the IR beam, record the spectrum, then return the widow to its orientation for spray-on. There is no need to interrupt the matrix gas flow during this operation. With acetone, an absorption at about 1720 cm^{-1} should be visible in the spectrum fairly soon after deposition begins.

131

Protocol 7 *Continued*

   (d) Open T3 to return the full volume of matrix gas mixture to the spray-on line.

   (e) Continue depositing the matrix, monitoring the progress of deposition by:

      i. recording IR spectra

      ii. noting the pressure drop on the Bourdon gauge (T3 can be temporarily closed again for this purpose), and

      iii. visually inspecting the cold window, if possible.

   (f) Adjust the needle valve as necessary, and continue deposition until the strongest IR band (that at 1720 cm^{-1} with acetone) gives about 10–30% transmission at its peak (absorbance = 0.5–1), then stop deposition by closing the needle valve. The total amount of gas mixture needed to reach this point will depend on the chosen matrix ratio. Ideally, for this test with acetone, spray-on should take between 30 minutes and two hours, but experience is needed to judge deposition times with any accuracy.

12. Make sure the cold window is in the optimum position in the IR beam, then record the IR spectrum.

13. Check the matrix visually. A good matrix will appear as a clear, even glass, without cracks or cloudiness, and may be almost invisible. The presence of cracks, lumps, or turbidity are signs of imperfect deposition, possibly due to a poor quality or dirty cold window, or to over-rapid deposition.

14. For an acetone–N_2 mixture used as a test of the matrix equipment, this is the end of the experiment. If it is desired, however, the experiment could be continued by depositing more of the matrix gas mixture, or by warming and re-cooling the matrix (annealing) to see if any molecular aggregation can be detected in the IR spectrum. In experiments with a photolabile guest species, the matrix can now be irradiated and a series of IR spectra recorded after each period of photolysis or annealing.

15. At the end of the experiment, get rid of the matrix by switching off the refrigerator and allowing the cold window to warm naturally. It is not usually necessary to use a heater. Observe the matrix as it evaporates. It may fall off the cold window in one piece and flutter about the bottom of the cold cell, or it may simply disappear smoothly. In either case, note the pressure rise in the cold cell. With the quantities of gas involved in the average matrix experiment, the pumping system should cope with the evaporation rate without problems.

16. Remove the gas bulb from the spray-on line, evacuate the line, and leave it under vacuum ready for the next experiment.

Protocol 7 provides a general method for depositing a pure gas or mixture of gases as a low-temperature matrix. If pulsed deposition is preferred, the procedure is exactly the same, except that the needle valve (NV in Fig. 4.5) is replaced by a pulsing volume (PV) between two taps (T7 and T8), and matrix deposition occurs in a series of fixed-volume pulses, rather than as a continuous stream (see Section 3.1.2). With the appropriate change of equipment, the matrices can be monitored by UV–visible or other forms of spectroscopy, rather than by IR.

The method given in the protocol for estimating the rate of deposition, using the Bourdon gauge, may seem a bit slow and cumbersome, but it will work in all circumstances. On some matrix, systems, however, it may be possible to monitor the rate of gas flow into the vacuum shroud more conveniently by observing the response of a Penning gauge mounted close to the shroud (cf. Figs 2.13 and 2.14). In several matrix-isolation systems which the author has used, spraying on gas at typical rates gave a response on the Penning gauge, and, with experience, the rate of spray-on could be adjusted to give pressure readings which correlated with the required rates of gas flow. The level of response of the Penning gauge depends on the efficiency with which gas is deposited on the cold window, which in turn depends on the size and geometry of the various components of the cold cell. In the author's present matrix systems, no Penning response is discernible at normal rates of spray-on, presumably because gas deposition is efficient; the use of the Penning gauge cannot therefore be recommended as a generally reliable way of monitoring the spray-on rate.

3.2 Evaporating a matrix at the end of an experiment

At the end of each experiment, the matrix has to be removed from the cold window. This can be done in three ways:

(a) The refrigerator is switched off and the matrix allowed to warm naturally.

(b) The refrigerator is kept running and the heater and temperature controller are used to warm the matrix slowly, until evaporation takes place. This method can be used to retard the rate of evaporation.

(c) The refrigerator is switched off and the heater used to boost the rate of evaporation.

The last of these is not recommended, except in special circumstances, because it is very easy to burn out the small resistance heaters supplied with the refrigerator. Over-vigorous evaporation can also splutter the matrix over the interior of the shroud—the vacuum chamber of the cold cell—and thus contaminate the inside surfaces of the external windows.

When matrices containing toxic materials are evaporated, the pumps should be vented into a fume cupboard or directly to the outside of the building. The possibility that a matrix might contain volatile materials that could damage the pumps or be otherwise harmful to the system should always be considered, but problems of this type seldom arise.

During evaporation of the matrix, it is best to maintain 'Dewar vacuum' inside the shroud. The pressure should not be allowed to rise above about 10^{-3} mbar, except for very brief periods (a few seconds at most). If the rate of evaporation exceeds the capacity of the pumping system to maintain this level of vacuum, the outside surfaces of the shroud, including the external windows, will be cooled and may condense atmospheric moisture. With KBr windows, fogging or even cracking might result. Over a period, it will become apparent which is the best way of removing matrices in a particular matrix system.

The occasional need to collect matrix products has been discussed in Chapter 1 (Section 2.2.2). In cases where the products are stable and sufficiently volatile, this may be achieved by pumping the evaporating matrix through a Pyrex or metal trap, fitted to the vacuum chamber of the cold cell, and cooled with liquid nitrogen. In other cases, an involatile product may be left behind on the cold window after the host gas has evaporated. Any such solid which survives warming to room temperature in the vacuum of the cold cell can be removed from the window with the aid of a small spatula (take care not to scratch the window) or by dipping the window and its holder into a small beaker of solvent, to dissolve the residue.

3.3 Less volatile guest species

When a guest material is not sufficiently volatile to be made up with the host gas as a mixture of known matrix ratio, it is necessary to evaporate it directly on to the cold window from a side-arm attached to the vacuum shroud of the cold cell, while simultaneously depositing an excess of the host gas. The exact requirements for volatilizing a relatively involatile guest vary widely. The least volatile materials may require very strong heating.

For many materials, a glass side-arm, in which the sample can be cooled in ice or a slush bath or be warmed to 200°C or so, will provide a sufficient range of spray-on conditions. Figure 4.6 shows a versatile side-arm of a type used frequently in the author's laboratory. The sample is placed in a small bulb between two vacuum taps. The bulb can be sized and shaped in various ways, but it should be possible to cool the bottom of it in a shallow bath. Alternatively, the sample can be warmed with a heat gun or by wrapping a small coil of heating wire around the bulb. The sample temperature can be measured by taping a thermometer or thermocouple to the outside of the bulb, with an overlayer of thermally insulating material such as glass wool.

In use, the side-arm is attached directly to the vacuum shroud of the cold cell at one end, and to the spray-on line at the other, via a short length of flexible stainless-steel tubing. In the author's laboratory, the side arm has standard 14/23 ground-glass cones at each end, which are waxed into aluminium adapters, to provide NW10 flange couplings. After the side-arm has been attached to the vacuum shroud and spray-on line, the sample is introduced via

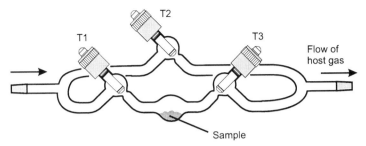

Fig. 4.6 A glass side-arm for less volatile guest materials.

T1 or T3 (Fig. 4.6), with a small spatula (solid) or a teat pipette (liquid). Very air-sensitive samples can be placed in the sample bulb in a glove box before being attached to the cold cell; the sample will then remain under an inert atmosphere, between T1 and T3, until the side-arm is evacuated. Photolabile samples should be protected by wrapping the sample bulb with aluminium foil. Once the side-arm containing the sample is in place, all three taps (T1–T3) are closed and the shroud and spray-on line evacuated as normal. Provided the spray-on line can be evacuated independently of the cold cell, and provided T1, T2 and T3 are kept closed, the cold window can be cooled at this stage.

Before matrix deposition can commence, the remaining air (or inert gas) in the side-arm must be evacuated through the spray-on line. The needle valve on the spray-on line should be fully opened for this. In evacuating the sample bulb via T1, great care should be taken to avoid losing all the sample down the pumps. For very volatile samples, it may be necessary to cool the sample to reduce its vapour pressure. In any case, it will probably not be possible to pump the sample bulb down to the normal low pressure of the vacuum system, and judgement will be needed to decide when the measured pressure corresponds with the vapour pressure of the sample. In this process, it will usually be necessary expose the Pirani gauge on the spray-on line to a low pressure of sample vapour. Try to limit this exposure to very brief periods. Needless to say, if there is the slightest chance of the Pirani gauge initiating an explosion of the sample vapour, any exposure of the gauge should be avoided; in this case ignorance (of sample vapour pressure) is preferable to injury.

The side-arm shown in Fig. 4.6 can cope with a wide range of sample volatility. For the most volatile samples, the host gas is allowed to flow into the vacuum shroud of the cold cell via T2. Gas flow is controlled by adjusting the needle valve on the spray-on line, as described in Protocol 7. It is a good idea to deposit pure host gas for a few minutes, before allowing any guest sample to evaporate. After this initial period, T3 is opened very slightly, and simultaneous deposition of host and guest begins. It may be necessary to cool the sample bulb to limit the rate of sample deposition. Deposition of the guest can be monitored by IR, UV–visible, or any other appropriate spectroscopic means. If the guest is evaporated too quickly, very concentrated matrices will result, with poor molecular isolation, and probably poor optical quality as well.

With very involatile samples, the host gas is passed directly over the sample via T1 and T3; T2 is kept closed. Passing the host gas over a sample in this way assists evaporation from the sample surface. The sample can also be heated, provided this does not result in decomposition. Below a certain level of volatility, however, heating the sample will merely tend to condense it on the cooler parts of the side-arm, down stream from the sample bulb. In this case, the sample will need to be evaporated from a point as close as possible to the cold window and with a direct ('line-of-sight') path to it There must be no vacuum taps, or even bends in the tubing, between the sample and the cold window. The simplest set-up consists of a short length (15–20 cm) of Pyrex tube (c. 15 mm o.d.) attached directly to the inlet port of the vacuum shroud and connected to the spray-on line at the other end. The sample is placed in the tube and remains open to the vacuum and the cold window throughout evacuation of the cold cell and cooling period. Provided it is sufficiently involatile, there will be no condensation of the sample on the cold window, until flow of the host gas and sample heating are begun. In extreme cases of involatility, a Knudsen cell with a furnace (see Chapter 3, Section 2.3) can be employed.

3.3.1 Matrix ratios with low-volatility samples

With guest species of low volatility, it is very difficult to measure matrix ratios. In many experiments, it may not be too important if the matrix ratio is not known, but it always best to obtain some idea of the concentration of the trapped species if possible. In order to estimate matrix ratios for matrices deposited by direct evaporation of the guest species and simultaneous deposition of the host gas, it is necessary to measure two separate quantities: the amount of guest material and the amount of host gas deposited on the cold window.

The number of millimoles of host gas allowed into the vacuum shroud of the cold cell is easily estimated from pressure readings on the spray-on line, but ascertaining the proportion of this gas which condenses on the cold window is another matter. The host gas will condense not only on the cold window but also on the window holder and other cold parts of the cold cell, and some will escape down the pumps. The proportion actually deposited on the window will vary from one matrix set-up to another. It is possible, however, to estimate the thickness of host gas accumulating on the cold window during deposition, by using He–Ne laser interference fringes.[5] Assuming the density of the solidified host gas is known, this leads to a measure of mmol h^{-1} actually deposited on the window. If a high quality micrometer needle valve is used to control the gas flow at reproducible rates and provided deposition conditions are standardized, the spray-on system can be calibrated in this way, and subsequently the number of millimoles of host gas deposited on the cold window can be estimated simply from the deposition time. As a matrix builds up, the

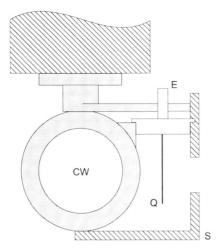

Fig. 4.7 A quartz-crystal microbalance mounted on the window holder of a cold cell. **CW**, cold window; **E**, electrical connections; **Q**, quartz crystal; **S**, copper shield.

efficiency of condensation of the host gas (or the *sticking coefficient*) may decrease, particularly owing to the inevitable temperature gradient within the matrix; and this will reduce the accuracy of the estimate of the number of millimoles of host in thicker matrices.

The amount of guest material deposited on the cold window can be estimated by using a quartz-crystal microbalance. The resonant frequency of a quartz crystal is lowered in proportion to the amount of material deposited on the resonator surface, and this effect has been widely used in microweighing techniques. The way in which such a microbalance can be adapted for matrix studies has been described by Werner Klotzbücher.[6] Figure 4.7 gives a simplified view of how the quartz crystal can be mounted close to the cold window and cooled to the same temperature. The quartz crystal is mounted on the window holder, perpendicular to the cold window. There is a copper shield, with a 10 mm diameter hole, around the outer side.

In use, the cold end is rotated so that the guest species can pass through the hole in the copper shield and condense on the quartz crystal. The desired rate of guest evaporation is first established by depositing on the crystal, then the flow of host gas is begun, and the matrix deposited by rotating the cold window to face the inlet port and source of guest species.

The use of a quartz-crystal microbalance can give reasonably good estimates of the matrix ratio in experiments where the guest and host are deposited quite separately. The method does not seem to be applicable when the host gas is passed over the guest material in a side-arm such as that shown in Fig. 4.6.

Quartz-crystal microbalances seem to have found favour in only a minority of matrix laboratories. This is because absolute values of the matrix ratio are not needed in a wide range of matrix studies, so there is little justification for the extra equipment costs and the added complexity of matrix deposition involved in using a microbalance.

The author's group recently made a study of the matrix photolysis of 4-hydroxyphenyl azides, in which varying the matrix ratio played a key role.[7,8] Photolysis of isolated molecules of these azides gave hydroxy-azacycloheptatetraenes (Scheme 4.1), but molecular aggregates of the same azides gave only azepin-4-ones. The azepinones are merely tautomers of the azacycloheptatetraenes, and are apparently formed by intermolecular proton transfer within aggregates of the initial photoproducts. Although it was not possible to estimate the matrix ratios in these experiments, it was straightforward to prepare either relatively dilute or relatively concentrated matrices, simply by varying the spray-on conditions, particularly the flow of host gas. This qualitative variation of matrix ratio will no doubt be adequate in other investigations as well.

Photolysis in N_2 matrices at 12–14 K:

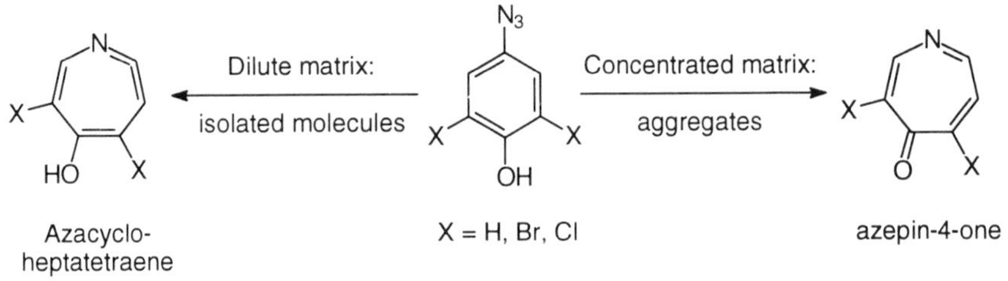

| Azacyclo-heptatetraene | X = H, Br, Cl | azepin-4-one |

Scheme 4.1

3.4 Elaborations of matrix spray-on techniques

The procedures for depositing matrices which have been described in previous sections are widely applicable and can be varied and elaborated endlessly. Once the basic techniques of slow spray-on and pulse deposition have been assimilated, workers new to matrix isolation will find numerous examples in the literature of specific adaptation of these methods to particular research areas. Two main elaborations of technique are worth mentioning briefly:

(a) the control of sample temperature during spray-on and

(b) the combination of spray-on with the generation of reactive species.

3.4.1 Control of sample temperature

For relatively involatile guest materials, careful control of the sample temperature during spray-on is often needed. In these cases, use can be made of:

• slush baths for cooling

• thermostatically controlled heating blocks

• Peltier-effect devices for both cooling and heating.

The normal type of laboratory recirculating constant-temperature baths, which can be equipped with cooling coils as well as heating elements, could also be used in this context, but they are both costly and bulky.

i. Slush baths

Slush baths are solid–liquid mixtures of solvents at their melting points, and are readily made up by carefully dropping small lumps of solid CO_2 into the solvent (down to $-78.5°C$) or by carefully pouring in liquid nitrogen. The temperature is maintained by ensuring that there is always some of the cryogen present. A slush bath in a shallow container can be used to cool a sample in a side-arm of the type shown in Fig. 4.6.

For this application the most useful slushes are likely to be those with temperatures in the range from just below room temperature down to about $-50°C$. Table 4.3 gives some suggestions, which provide reasonably good coverage from $+13$ to $-50°C$. Some of these solvents are quite toxic, such as carbon tetrachloride, most are inflammable, and, when handling solid CO_2 or liquid nitrogen, there is always a risk of frostbite; so care should be taken at all times when preparing, using and disposing of slush baths.

ii. Heating blocks

Small aluminium or copper blocks, drilled to take glass or metal, open-ended sample tubes, can be easily constructed and fitted with heating elements and

Table 4.3. Single-solvent constant-temperature slush baths

T / °C	Slush bath solvent	Principal known hazards
13	p-Xylene (1,4-dimethylbenzene)	Harmful, flammable
12	1,4-Dioxane	Irritant, flammable
7	Cyclohexane	Flammable
3	Formamide	Irritant
0	Ice-water	
−6	Aniline	Toxic
−8	Methyl salicylate	
−10	Diethylene glycol	Harmful by ingestion
−12	t-Amyl alcohol (2,2-dimethylpropan-1-ol)	Harmful, flammable
−12	Cycloheptane	Flammable
−13	Ethylene glycol (ethane-1,2-diol)	Harmful by ingestion
−14	Benzaldehyde	Harmful by ingestion
−15	Benzyl alcohol	Harmful
−17	o-Dichlorobenzene	Harmful
−23	Carbon tetrachloride	Toxic
−29	o-Xylene (1,2-dimethylbenzene)	Harmful, flammable
−30	m-Toluidine (3-methylaniline)	Toxic
−38	Thiophene	Eye irritant
−41	Acetonitrile	Toxic, flammable
−45	Chlorobenzene	Harmful, flammable
−47	m-Xylene (1,3-dimethylbenzene)	Harmful, flammable
−50	Diethyl malonate	Irritant

thermocouples. If the vacuum shroud of the cold cell is large enough, a heating block of this type can be mounted inside the shroud. The temperature of the block is measured with a thermocouple and controlled either manually or automatically. This type of spray-on aid works only with samples which have negligible vapour pressure at room temperature, because the sample must be loaded into the block before the cold cell is evacuated and cooled.

iii. Peltier-effect heat pumps
Peltier-effect heat pumps are semiconductor thermoelectric devices, supplied in the form of square modules, a few centimetres on each side and a few millimetres thick. When a DC current is passed through the device, a thermal gradient is set-up, such that one side of the device is warmed while the other is cooled. The temperature differential increases with increasing current, up to a practical maximum of about 50–60°C, and is reversed when the current direction is reversed. The advantage of these devices is that both cooling and heating can be achieved electrically and with a very compact set-up.

Figure 4.8 shows a simple spray-on aid which uses a Peltier-effect device to provide sample heating and cooling. One side of the device is in thermal contact with the sample, while the other is maintained at constant temperature by a flow of water. With a water temperature of about 10°C, sample temperatures in the range of approximately −30 to +50°C can be achieved. With this type of spray-on aid, the sample is maintained at low temperature while the cold cell is evacuated and cooled, and is subsequently warmed for matrix deposition. The rate of deposition can be controlled by adjusting the current through the Peltier device.

3.4.2 Combined generation of reactive species and spray-on
There are numerous ways in which reactive species can be generated in the gas phase and then trapped in low-temperature matrices. Some of these have been

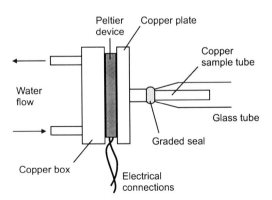

Fig. 4.8 A spray-on aid using Peltier-effect heating and cooling. The sample is placed in a small copper tube in thermal contact with the Peltier device.

discussed in Chapters 1 and 3. The techniques employed include vacuum pyrolysis, gas-phase photolysis, microwave discharge, and chemical reaction.

In most cases, the matrices are deposited by the slow spray-on method. For example, a mixture of host gas and guest can be passed through a pyrolysis tube, such as that shown in Fig. 3.4, and the resulting pyrolysate deposited as the matrix. The procedure of Protocol 7 for slow matrix deposition can be followed exactly in this type of experiment. Alternatively, for less volatile guest materials, a side-arm like that shown in Fig. 4.6 can be attached to the inlet end of the pyrolysis tube. The pulsed pyrolysis technique (see Chapter 3, Section 2.2) is a rare example of the gas-phase generation of reactive species combined with pulsed matrix deposition.

3.5 Window cleaning

In a high proportion of matrix experiments, a deposit is left on the cold window, which therefore has to be cleaned. The time needed to warm the cold cell to room temperature, clean the window, and restore the vacuum in the shroud to an acceptable level (10^{-4} mbar or better) determines the rate at which matrix experiments can be carried out. It is seldom possible to exceed one experiment per day. For very exacting studies, where, for example rigorous exclusion of traces of water vapour is essential, pumping of the vacuum shroud after window cleaning may need several days. The best way of cleaning the cold window and other interior parts of the cold cell has been discussed in Chapter 2 (Section 8.2.3).

References

1. Meyer, B. *Low Temperature Spectroscopy*; Elsevier: New York, **1971**.
2. Hallam, H. E., ed. *Vibrational Spectroscopy of Trapped Species*; Wiley: London, **1973**.
3. Cradock, S.; Hinchcliffe, A. J. *Matrix Isolation*; Cambridge University Press, **1975**.
4. Moskovits, M.; Ozin, G., ed. *Cryochemistry*; Wiley: New York, **1976**.
5. Groner, P.; Stolkin, I.; Günthard, H. H. *J. Phys. E* **1973**, *6*, 122–123.
6. Klotzbücher, W. E. *Cryogenics* **1983**, *23*, 554–556.
7. Dunkin, I. R.; El Ayeb, A. A.; Lynch, M. A. *J. Chem. Soc., Chem. Commun.* **1994**, 1695–1696.
8. Dunkin, I. R.; El Ayeb, A. A.; Gallivan, S. L.; Lynch, M. A. *J. Chem. Soc., Perkin Trans. 2*, **1997**, 1419–1427.

Matrix photochemistry and spectroscopy with plane-polarized light

This chapter describes matrix experiments with plane-polarized light, in which molecules can be spatially oriented. Although not a major part of matrix photochemistry, the experiments are easily carried out, and can provide useful information about the symmetries of electronic and vibrational transitions and about molecular conformations. The experiments depend on the fact that, except for very small molecules, trapped species in matrices are not only prevented from diffusing but also from rotating.

A key molecular property in these experiments is the directionality of transition moments. This can be—and usually is—overlooked in the majority of photochemical and spectroscopic studies, where molecules are freely rotating or randomly oriented. It comes to the fore, however, with solid-state samples in which the molecules are aligned, such as crystals. Since this and other related concepts may be unfamiliar to some readers, the relevant background theory will be discussed briefly in the next section. Recommended conventions for the relevant terminology are given in Section 1.6.

1. Photoselection and linear dichroism

1.1 Transition moments

When a molecule absorbs a photon and becomes electronically excited it undergoes a transition from one electronic energy level—usually the ground state—to another, higher level. This is brought about by interaction of the oscillating electric field of the light (the E-vector) with the electrons in the molecule. Every electronic transition has an associated *transition moment*. For molecular excitation, the transition moment gives the probability of absorption of a photon of a particular frequency. It is a vector quantity and hence there is a direction within the molecule with which the E-vector must be aligned for maximum probability of absorption.

If the molecules of a sample are fixed in space and cannot rotate, and if electronic excitation is induced with plane-polarized light, there will be some

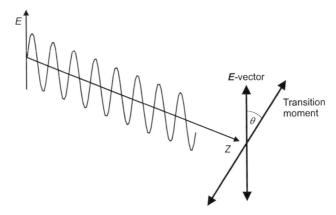

Fig. 5.1 Absorption of polarized light. A beam of plane-polarized light propagating in the Z-direction with the plane of polarization vertical. The probability that a molecule will absorb a photon is proportional to the angle, θ, between the vertical *E*-vector and the molecular transition moment.

molecules oriented with their transition moments parallel or nearly parallel to the *E*-vector, and these will have a high probability of absorption, while others will be oriented with their transition moments perpendicular or nearly perpendicular to the *E*-vector, and these will have a low probability of absorption. To be precise, the probability that a non-rotating molecule will absorb a photon of the right frequency is proportional to $\cos^2\theta$, where θ is the angle between the *E*-vector of the light and the transition moment of the relevant electronic transition moment of the molecule (Fig. 5.1).

In molecules with high symmetry (C_{3v}, C_{4v}, octahedral, etc.), degeneracies exist which can result in equal probability of absorption for any direction within a plane in a molecule, or even over the entire sphere. For the majority of molecules, however, with lower symmetry (C_{2v} or less), each transition moment is confined to a single direction.

Besides electronic excitation, other types of light-induced transitions between molecular energy levels also have their associated transition moments. In particular, each vibrational transition of a molecule has an associated direction within the molecular framework, with which the *E*-vector of IR radiation of the right frequency must be aligned for maximum probability of absorption. This fact can be exploited experimentally, as shown in the following sections.

1.2 Photoselection

When molecules trapped in rigid matrices are photolysed with plane-polarized light, those with their electronic transition moments aligned parallel to the *E*-vector of the light are photolysed preferentially. This is called *photoselection*.

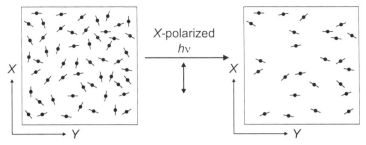

Fig. 5.2 Photoselection of molecules trapped in a rigid matrix. Each molecule is represented by a solid circle and a line indicating the direction of the transition moment of its photoactive absorption. Photolysis with *X*-polarized light preferentially removes molecules whose transition moments lie close to the *X*-direction. this leaves an anisotropic sample containing molecules of starting material with a net preferred orientation perpendicular to the *X*-direction. *(Adapted with permission from ref. 21. © 1997, Elsevier.)*

Except with highly symmetrical molecules, where degeneracies can occur, partial photolysis of a matrix sample with plane-polarized light will produce an array of residual starting-material molecules with a net preferred orientation (Fig. 5.2). The photoproduct generated in this way will also be preferentially aligned, provided that thermal randomization of the direction of photoproduct molecules does not occur while they are losing their excess energy to the surrounding matrix. The overall result is that a matrix which began as an isotropic sample will become anisotropic.

Interest in molecular alignment and the associated optical phenomena goes back a long way. In 1919, Fritz Weigert observed that polarized photolysis produced an anisotropic spot on photographic paper,[1] and concluded that there existed electric vectors linked with absorptions which were dependent on the orientation of each individual molecule. He subsequently photolysed a variety of amorphous isotropic materials, such as dyes in gelatin,[2,3] and extended these studies to immobilized biological extracts and anisotropy in retinal absorptions.[4,5] In 1943, Gilbert Lewis used polarized light in his studies of photo-oxidation in frozen organic glasses (see Chapter 1) and achieved the generation of preferentially oriented product molecules.[6] Work on photoselection in frozen organic glasses was continued by Andreas Albrecht, who also reviewed the technique and its theory.[7–9]

1.3 Linear dichroism

The generation of an anisotropic sample, in which the molecules have a preferred alignment, shows up spectroscopically as *linear dichroism*. This means that a given spectroscopic absorption has a different intensity when it is observed under different polarizations. For example, in Fig. 5.2, the starting-material molecules left after *X*-polarized photolysis have a net preferred

orientation with their electronic transition moments in the laboratory
Y-direction. The intensity of the absorption associated with this transition
moment, that is the *photoactive* absorption which led to photolysis, will show a
maximum when measured with Y-polarized light and a minimum when
measured with X-polarized light. In other words, the absorbance measured
with Y-polarized light (A_Y) will be greater than that measured with
X-polarized light (A_X).

Of necessity, the photo-active absorption will show linear dichroism in this
sense, i.e. $A_Y > A_X$ after X-polarized photolysis, or $A_X > A_Y$ after Y-polarized
photolysis. In the author's laboratory this is referred to as *parallel linear
dichroism* or *parallel polarization* of the absorption band. With other
absorptions, however, linear dichroism in the opposite direction can be
observed, i.e with $A_X > A_Y$ after X-polarized photolysis, or $A_Y > A_X$ after Y-
polarized photolysis. This is called *perpendicular linear dichroism* or
perpendicular polarization. It arises, for example, where a transition moment
for an absorption is perpendicular to that of the photo-active transition.

The extent to which orientation has been achieved is indicated by the
dichroic ratio (A_X/A_Y) or by the *degree of polarization*, P, where

$$P = \frac{A_Y - A_X}{A_Y - A_X}.$$

Figure 5.3 shows UV spectra recorded during the polarized photolysis of
diazocyclopentadiene in a CO matrix.[10] The product in the reaction was the

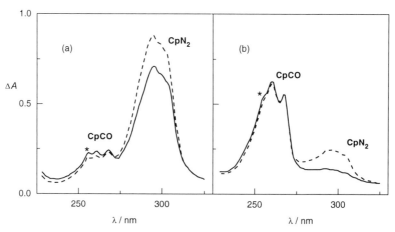

Fig. 5.3 Linearly dichroic UV absorption spectra. Diazocyclopentadiene (Scheme 5.1,
CpN$_2$) in a CO matrix at 12 K (CO:CpN$_2$ = 1000:1): (a) after 3 minutes photolysis with X-
polarized light (λ = 315 ± 10 nm); (b) the same matrix after 27 minutes of X-polarized
photolysis. In each case two spectra were recorded: X-polarized (solid line) and Y-
polarized (broken line). The product was the ketene, CpCO. The small band at 256 nm
marked with an asterisk is an absorption of the CsBr cold window. *(Adapted with
permission from ref. 10. © 1981, American Chemical Society.)*

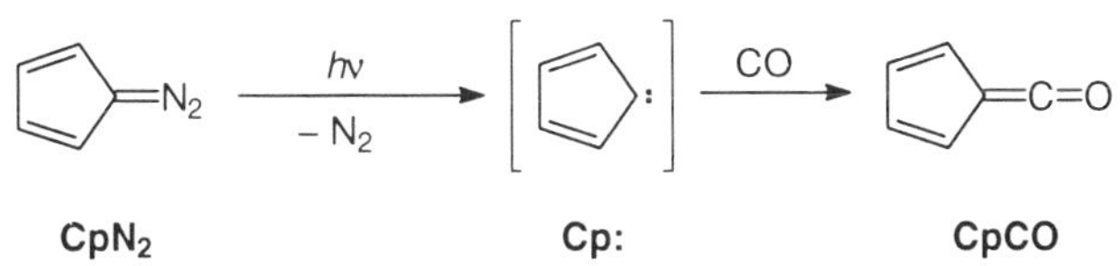

Scheme 5.1

corresponding ketene (Scheme 5.1). Both the starting diazo compound and the product exhibited the expected linear dichroism. The product, however, had a much lower degree of polarization than the residual starting material, suggesting that some molecular rotation occurred during the reaction, but not enough to ensure complete randomization of product orientation.

Both the electronic and vibrational absorptions of a partially aligned array of molecules will exhibit linear dichroism, the sense of which will depend on the directions of the individual transition moments in the molecules. The sense of polarization—parallel or perpendicular—is often given a relative sign and thus denoted as positive or negative. The absolute sign, that is, which sense of polarization is positive and which negative, is arbitrary, however.

Photoselection is not the only way to achieve partial alignment of solvent molecules in solid or glassy hosts. The most common alternative method involves the stretching of polymer films, but orientation by electric fields and within liquid crystals has also been investigated. For low-temperature matrices, however, photoselection appears to be the only available method. Application of the technique to solidified gas matrices brings the advantage that IR dichroism can be studied throughout the mid-IR range. The first studies of this kind were made on metal carbonyls[11] and diazocyclo-pentadiene[10] by Jim Turner's group. The theory of such dichroism has been discussed previously by several authors,[7–9,11–17] and is covered very fully in a book by Josef Michl and Erik Thulstrup.[18]

1.4 'Invisible' molecules

Figure 5.2, which illustrates selective photolysis with plane-polarized light, is oversimplified in one respect: it neglects molecules which have their transition moments parallel or nearly parallel to the Z-direction, that is, the beam direction. Assuming that the beam is perfectly collimated, molecules with their transition moments parallel to the beam direction will not be able to absorb light, because their transition moments will be perpendicular to the E-vector. This is true regardless of whether the beam is polarized and applies equally for photolysis and spectroscopic analysis.

It follows that, during a normal matrix photolysis with unpolarized light, the last molecules of starting material to be photolysed will tend to be those aligned in the beam direction, since these molecules have the lowest probability of absorption. Thus even with unpolarized light, photolysis will produce an anisotropic sample, with molecules of residual starting material having a net preferred orientation of their transition moments in the beam direction. After nearly complete conversion, there may still be substantial numbers of starting-material molecules in the beam direction, but, at the same time, the measured absorbance at λ_{max} of the photo-active transition might be close to zero. The residual molecules are thus 'invisible' to the analysing beam. This is a startling conclusion when first encountered.

The situation with the vibrational spectrum of such a sample is more complicated. Vibrational bands with transition moments parallel or nearly parallel to the transition moment of the photo-active absorption will behave in the same way, that is, they will tend to become 'invisible' as photolysis progresses. On the other hand, vibrational bands with transition moments perpendicular to that of the photoactive absorption will develop enhanced relative intensities as photolysis progresses. At nearly complete conversion, the remaining molecules of starting material will be preferentially aligned with their photo-active transition moments parallel to the beam direction, while the transition moments of vibrational transitions perpendicular to that of the photoactive transition will be preferentially oriented in the XY-plane, that is, perpendicular to the beam direction. Again, this is true regardless of whether the photolysis or spectrometer beams are polarized.

In general, therefore, during photolysis of molecules trapped in a rigid matrix, the decrease in absorbance of electronic and vibrational absorptions will not be proportional to the numbers of molecules photolysed. In actual experiments, Gilbert Lewis was able to observe the preferential orientation of molecules in frozen organic glasses following unpolarized photolysis,[6] but it has so far been difficult to demonstrate these effects in the much thinner solidified gas matrices. Imperfect beam collimation and scattering, reflection or refraction of light within the matrix will all increase the probability that molecules lying in the Z direction will absorb and be photolysed. Thus, in practice, the complete photolysis of the starting-material can be achieved in a matrix experiment—something of a relief in most circumstances. Nevertheless, anyone carrying out matrix-photolysis experiments should be aware of the possibility of these orientational effects.

1.5 Photo-reorientation

The process of photoselection described in the previous sections involves the preferential destruction of X- or Y-oriented molecules of the starting material. In some systems, however, polarized irradiation of the sample can result in molecular rotation, without loss of the starting material. The ketene, CpCO (Scheme 5.1), provides a good example.[10]

Figure 5.4 shows UV absorptions of this ketene, which was generated by unpolarized photolysis of diazocyclopentadiene (CpN$_2$, Scheme 5.1) in a CO matrix. The photolysis was carried out to completion (cf. Fig. 5.3). At this stage the CpCO molecules were randomly oriented, and the X- and Y-polarized UV

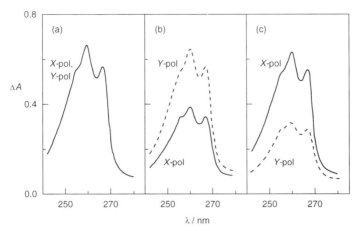

Fig. 5.4 Photo-reorientation of the ketene, CpCO (Scheme 5.1). (a) CpCO in an N_2 matrix at 12 K, generated by complete (unpolarized) photolysis of CpN_2 (N_2:CpN_2 = 1000:1) with λ = 315 ± 10 nm; the X- and Y-polarized spectra (X-pol, Y-pol) were completely superimposable. (b) The same matrix after 15 hours X-polarized photolysis (λ = 257 ± 10 nm), and (c) after a further 71 hours photolysis with Y-polarized light (λ = 257 ± 10 nm). *(Adapted with permission from ref. 10. © 1981, American Chemical Society.)*

spectra were identical (Fig. 5.4(a)). Subsequent irradiation with X-polarized light generated a matrix in which the CpCO molecules were preferentially oriented in the Y direction (Fig. 5.4(b)). When the polarization of the UV light was switched from X to Y, further polarized irradiation gave a matrix in which the CpCO molecules were preferentially oriented in the X direction (Fig. 5.4(c)). The change of molecular orientation was accomplished with very little loss of overall intensity. It is quite clear from the increase in X-polarized absorption between Fig. 5.4(b) and Fig. 5.4(c) that molecules had been re-oriented by the polarized UV light. This experiment was continued for several more days, during which time the preferred orientation of the ketene molecules was reversed once again, with no appreciable loss of overall intensity. Thus, up to a point, the ketene molecules cold be 'shepherded' from one orientation to another.

This photoinduced molecular reorientation could have occurred as the result of either a photophysical or a photochemical process. It is possible that excited-state ketene molecules simply lost their excess energy to the surrounding host molecules through internal conversion, thus imparting a good deal of vibrational energy to the matrix. During this process, any temporary softening or melting of the local matrix would have allowed rotation of the trapped ketene molecules. It is more likely, however, that photochemistry was involved. A CpCO molecule is shaped rather like a tennis racket, with a flat ring and a long handle in the form of the C=C=O moiety, and this handle would severely impede rotation. It seems probable, therefore,

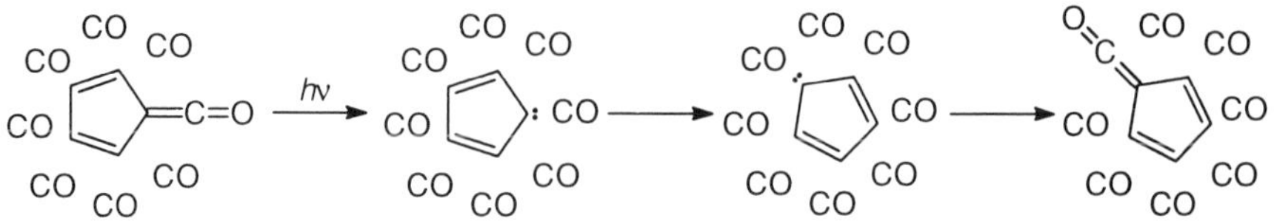

Scheme 5.2

that absorption of a photon by a CpCO molecule resulted in photolysis and ejection of CO (Scheme 5.2). The resulting carbene molecule (Cp:) is more nearly disk-like than CpCO, and could easily have rotated in its own plane while the excess vibrational energy was being dissipated to the matrix. Alternatively, the equivalent of carbene rotation could have occurred by sequential migration of hydrogen atoms around the ring. When the carbene molecule stopped rotating, by whatever mechanism, it simply reacted with the nearest CO molecule to regenerate CpCO. Remember that the experiment was conducted with CO as the host species.

Whether reorientation occurred photophysically or photochemically, the driving force was the same. Molecules which were closely aligned with the **E**-vector of the UV light preferentially absorbed photons and dissociated into CO and Cp: pairs. The carbene molecules thus generated rotated at random and finally reacted with CO molecules to regenerate CpCO. It was a matter of chance which orientations the resulting CpCO molecules adopted, but molecules which ended up approximately perpendicular to the direction of the **E**-vector of the UV light had a very low probability of reabsorption, while those that ended up aligned close to the **E**-vector could reabsorb and undergo the whole process again. Over a period, therefore, molecules accumulated in orientations with a low probability of photon absorption. This is an example of Le Chatelier's principle: that a system will react to an external influence in such a way as to minimize the effect of that influence.

There is one last point about the spectra shown in Fig. 5.4 that needs to be cleared up. In the initial period of X-polarized photolysis, molecules were apparently lost (cf. Fig. 5.4(a) and (b)). The X-polarized absorption dropped to less than 60% of the initial value, but there was no increase, rather a slight decrease, in the Y-polarized absorption. Thus the total absorption by ketene molecules decreased and this would normally be taken as an indication of photolysis. In view of the subsequent behaviour of the sample under pro-longed UV irradiation, however, it is clear that photolysis, that is, the removal of ketene, did not take place. Therefore the only explanation of the initial loss of absorbance is that the CpCO molecules were being re-oriented from the X direction into both the Y and Z directions. As explained in the preceding section, the Z-oriented molecules (parallel or nearly parallel to the beam direction) would have been 'invisible' to the UV spectrometer. This re-orientation into the Z direction clearly reached a limit, however, because no significant apparent loss of the ketene occurred during the remainder of the experiment, which involved very prolonged UV irradiation. This observed limit to Z reorientation may have resulted from the fact that a matrix-isolated molecule of the carbene intermediate (Cp:) could rotate only in its own plane and not tumble in all directions, thus limiting the extent of possible reorientation, but may simply have been due to imperfect collimation of the photolysis beam or light scattering within the matrix.

1.6 Recommended conventions

At this stage it may be helpful to consider the terms used to describe the alignment of molecules in experiments with polarized light. The literature on this subject contains the terms *photoselection, photo-orientation* and *photo-reorientation*, and in the early days there was a tendency to use the first two interchangeably. Nevertheless, there are three distinct ways in which an oriented array of molecules may be generated in experiments of this type, and it seems best to reserve each term for a specific use, as follows:

- *photoselection*—preferential photolysis leaving partially oriented starting material
- *photo-orientation*—photolytic generation of a partially oriented product
- *photo-reorientation*—light-induced reorientation of molecules without photolysis.

Note that photoselection of the starting material will also result in photo-orientation of the product, provided randomization by molecular rotation does not occur in the process. Moreover, if a photolytic reaction is partly reversible and partly irreversible, photo-reorientation of the starting material can occur simultaneously with photoselection and photo-orientation.

In addition to these terms used to describe molecular alignment, the following conventions are normally adopted when discussing polarized photolysis and molecular orientation.

(a) Laboratory axes are denoted with upper case letters: X, Y, Z.

(b) Molecular axes are denoted with lower case letters: x, y, z.

(c) In each case a set of right-handed axes is adopted (see Fig. 5.5).

(d) The beam direction for photolysis or spectroscopic analysis is the laboratory Z-axis.

(e) Unless there are good reasons for a different choice, the molecular z-axis is taken as the transition-moment direction for the photoactive transition.

1.7 Molecular symmetry

In molecules with at least some symmetry, the transition moments of the various electronic and vibrational transitions will not be able to take any direction within the molecular framework: the allowed directions will be

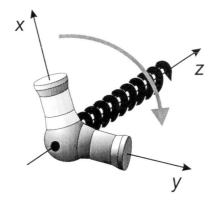

Fig. 5.5 Right-handed axes, defined according to the following convention. The handle of a right-handed corkscrew is bent to resemble x- and y-axes. Rotating the x-axis clockwise towards the y-axis, as shown, advances the corkscrew into the cork—along the z-axis.

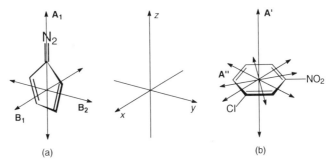

(a)　　　　　　　　　　　　　　　　(b)

Fig. 5.6 Allowed directions for transition moments. (a) A C_{2v} molecule, showing directions of A_1, B_1, and B_2 transition moments; (b) a C_s molecule, showing directions of A′ and A″ transition moments.

constrained by the molecular symmetry. The full range of possibilities are included in published point-group character tables,[19] but two simple examples will illustrate the idea. For molecules with C_{2v} symmetry, such as diazocyclopentadiene (Fig. 5.6(a)), every transition moment for allowed electronic and vibrational transitions must lie along the x-, y- or z-axis of the molecule. The transitions thus fall into three groups, with A_1 (z), B_1 (x), or B_2 (y) symmetry. The transition moments of each group are perpendicular to those of the other two groups. The transition moments of molecules of C_s symmetry, that is those with a single mirror plane, such as 3-chloro-1-nitrobenzene (Fig. 5.6b), fall into just two groups: A′ and A″. The A″ transition moments lie in one direction, the molecular z-direction, say, while the A′ transition moments can take any direction in the molecular xy plane. Thus all the A″ transition moments are perpendicular to all the A′ transition moments, but, within the A′ group, the various transition moments may lie at any angle to each other.

152

Comprehensive treatments of this subject can be found in Michl and Thulstrup's book[18] and in books on molecular symmetry and group theory.[20] Two recent papers from the author's group discuss problems of interpretation of linear dichroism with molecules of low symmetry, and provide references to previous work in this area.[21,22]

2. Experimental techniques for polarized photolysis and polarized spectroscopy of matrices

2.1 Polarizers

With a matrix-isolation cold cell, a suitable lamp, and one or two polarizers, experiments in the polarized photolysis of matrix samples can be carried out very easily. The experiments will require a polarizer for the photolysis beam and also one for any spectrometer with which it is intended to observe linear dichroism. The exact polarizers required will depend on the kind of experiments envisaged. Some of the available types are discussed in the next two sections. Because the polarized light needs to be collimated reasonably well, extended light sources, such as low-pressure Hg arcs are not as suitable as point sources such as high-pressure Hg arcs. The set-up shown in Fig. 3.3, a point source with a quartz collimating lens and water filter, is used in the author's laboratory for nearly all studies involving polarized photolysis. The polarizer is placed after the water filter, to prevent damage by overheating.

2.1.1 UV and visible polarizers

i. Plastic sheet and photographic polarizers

The plane-polarization of visible light can be achieved with a dichroic film polarizer in the form of a plastic sheet or a glass-mounted photographic polarizing filter. Such polarizers can be bought from many laboratory and photographic suppliers and provide an inexpensive option for preliminary studies. Typical sheet polarizers transmit useful amounts of light down to about 400 nm (Fig. 5.7(a)), while photographic polarizing filters extend this range down to about 330 nm (Fig. 5.7(b)). When the polarizers are in good condition, only about 0.1–1% of light of the 'wrong' polarization will be transmitted across the usable wavelength range.

These types of polarizer can be used for photolysis or for polarizing the beam of a UV–visible spectrometer, but they are not designed to cope with high intensities, and will inevitably deteriorate through long exposure to UV light. They are also prone to damage by overheating. Therefore, when they are used with high intensity light in photolysis, they should always be placed in the optical train after a water filter or a monochromator, so that most of the IR radiation is eliminated. In any case, the performance of the polarizers should be checked regularly, and, with the low cost of replacement, detectable deterioration need not be tolerated.

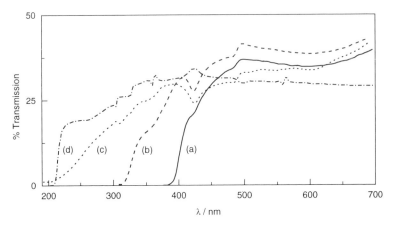

Fig. 5.7 UV–visible spectra of polarizers. (a) Plastic sheet polarizer; (b) photographic glass polarizing filter; (c) dichroic film polarizer on a quartz substrate; (d) Glan–Taylor prism. Each spectrum was recorded with the plane of polarization at approximately 45° to the laboratory vertical axis (see Section 2.1.4).

Plastic sheet polarizers can be cut into various shapes and sizes. The sheets are usually purchased as rectangles with sides of about 10–20 cm; and the direction of maximum transmission of plane-polarized light will usually be parallel to the longer sides. When cutting the sheet, this direction should be noted and marked on the edge of the cut piece in some way. If this precaution is omitted, it is possible to redetermine the direction of polarization by measuring the transmission with respect to a second polarizer of known orientation, but some error is bound to creep in with circular cut-outs.

Photographic polarizing filters are supplied in rotating mounts and there is usually a mark on the rim of the mount to indicate the plane of polarization. Confusingly, some filters have this mark positioned to show the direction perpendicular to the plane of polarization; so the actual plane of polarization should be checked against another polarizer of known orientation, before the filter is used.

ii. Dichroic film polarizers on quartz substrates

A more extended wavelength range in the UV than is provided by plastic sheet or photographic polarizers can be obtained by use of dichroic film polarizers applied as coatings on quartz substrates. These are usually supplied as discs of various diameters, with the direction of polarization indicated by marks near the edges. Polarizers of 50 mm diameter are the best choice for both photolysis and spectrometer polarization.

Film polarizers on quartz discs transmit useful amounts of UV light down to about 250 nm (Fig. 5.7(c)), but otherwise have a similar specification to plastic sheet or photographic polarizers. They are equally prone to damage through overheating and prolonged exposure to UV light, but are unfortunately much more expensive to replace. In polarized photolysis, therefore, it is best if they can be used with monochromated light, to minimize the total incident intensity. In many experiments, however, it will be found that mono-chromation of the light will lead to excessive photolysis times; so a balance has to be struck.

For polarizing UV beams, dichroic film polarizers on quartz disks are probably the best choice in terms of cost and versatility, despite their disadvantages.

iii. Prism polarizers

Polarizers constructed from two or three calcite prisms outperform film polarizers in eliminating the unwanted polarization. *Extinction ratios* (i.e. ratio of unwanted to wanted polarization transmitted) lie typically in the range 10^{-4}–10^{-6}, compared with 10^{-2}–10^{-4} for film polarizers. Of the several designs of prism polarizers, the Glan–Taylor polarizer seems best suited for polarized photolysis in matrix studies, having a relatively small length to aperture ratio (0.85–1.0), and thus being relatively compact. These prisms also have better transmission of UV light than dichroic film polarizers, with more than 10% transmission down to about 220 nm (Fig. 5.7(d)).

Despite their high performance, prism polarizers suffer from several dis-advantages. Firstly, they are substantially more expensive than film polarizers, although they should deteriorate more slowly, provided a correctly specified polarizer is used for high-intensity light. Secondly, apertures greater than 20 mm are not generally available, and, although adequate, this is smaller than ideal for matrix studies. Thirdly, they are much bulkier than film polarizers. A typical Glan–Taylor polarizer of 20 mm aperture will have a total length of about 30 mm, including its mounting tube, and this can cause problems of location, especially in spectrometer sample compartments. Finally, whereas film polarizers are not very sensitive to the incident beam angle, prism polarizers work optimally over a fairly narrow cone, typically 8–16°. Thus, beam collimation and correct alignment of the polarizer need careful setting up. With some prism designs, two beams of opposite polarization emerge from the polarizer with only a small angle between them (5–20°); so that adequate beam separation is obtained only at some distance.

Where the superior performance of a prism polarizer seems necessary, care should be taken to ensure that the specification matches the intended use. Anti-reflection coatings should be avoided, because they reduce the useful UV range, while jointing materials between the prism elements reduce both the wavelength range and the intensity of light which can be safely transmitted. Designs for high-intensity UV transmission have air gaps between the elements. A polarizer designed for use with high-intensity UV light can, of course, be used for polarized UV–visible spectroscopy of matrices, provided there is enough room in the sample compartment to accommodate both the cold cell and the polarizer.

2.1.2 Infrared polarizers

Polarizers for the IR region consist of very fine aluminium grids deposited lithographically on various substrates, such as KRS-5 (thallium bromoiodide), CaF_2, germanium or polyethylene. Unfortunately, there are no inexpensive variants. For complete coverage of the mid-IR, a KRS-5 polarizer should be chosen (5000–300 cm^{-1}). Polyethylene polarizers give extended coverage in the far-IR (500–10 cm^{-1}). Transmission of the 'unwanted' polarization varies with frequency, but can be expected to lie in the range 0.25–4%.

IR polarizers are usually supplied as disks with usable apertures of about 25 mm, which is adequate for polarized matrix IR studies, and are typically 6–8 mm thick. They are thus reasonably compact and can normally be readily accommodated in the sample compartment along with the cold cell.

2.1.3 Polarizer mounts and positioning

It is important to be able to reproduce the orientation of a polarizer to within a degree or two, especially when recording polarized spectra. Although a makeshift mount, made from cardboard, for example, can be used, it is far preferable to install each polarizer in a rotating mount with an accurate degree scale engraved around the edge. Suitable mounts can be obtained from the suppliers of optical components, and can usually be ordered with the polarizers themselves.

For polarized photolysis, the polarizer should be placed as close as possible to the external window of the cold cell through which the incident beam passes. It can be mounted on the shroud, or independently as part of the optical train of the light source. Since reflection of a beam can alter its polarization, there should be no mirrors (to change beam direction) or other optical components between the polarizer and the cold cell.

For polarized spectroscopy, the polarizer can be inserted in the beam either before or after the cold cell, and it can be mounted either on the shroud or on the entrance or exit aperture of the sample compartment. As with photolysis, there should be no optical component, such as a mirror, between the polarizer and the cold cell.

In spectrometers with both sample and reference beams, one way to eliminate polarizer absorptions is to mount a polarizer in each beam. Apart from the added expense of two polarizers, this is not recommended, because both polarizers have to be realigned accurately at each change of polarization, and also because supposedly identical polarizers can differ significantly in their spectra. It is better to use a single polarizer, record a background spectrum for each of the polarizer orientations to be used in an experiment, and then subtract the appropriate background from each subsequent polarized spectrum.

2.2 Instrument polarization and the choice of laboratory axes

2.2.1 Beam polarization in spectrometers

The beams of most IR and UV–visible spectrometers are distinctly polarized. Although this does not give rise to problems in normal spectroscopy, and is seldom discussed, there are serious practical consequences when polarizers are to be used. The polarization is caused by various optical components of the spectrometers, particularly diffraction gratings.

Figure 5.8 shows two UV–visible spectra of a dichroic film polarizer, recorded on a fairly typical spectrometer, with the plane of polarization vertical in one case and horizontal in the other. The two spectra differ very markedly. At 700 nm the ratio of horizontally polarized to vertically polarized light is more than 4:1. The spectra cross over at several points, including the wavelength at which a change of lamps, from tungsten to deuterium, takes place. The spectrometer beam is clearly highly polarized over much of the wavelength range.

Grating IR spectrometers are similarly highly polarized, as shown in Fig. 5.9. In this example, the ratio of horizontally polarized to vertically polarized light just above 600 cm^{-1} is more than 8:1; the spectra cross at three points; and there are discontinuities where a change of grating order at 2000 cm^{-1} and a change of grating at 600 cm^{-1} occur. FT-IR spectrometers do not have gratings, but they can still be expected to show some beam polarization. Figure 5.10 shows this for a standard Nicolet model, although the degree of polarization is much less than for a grating instrument.

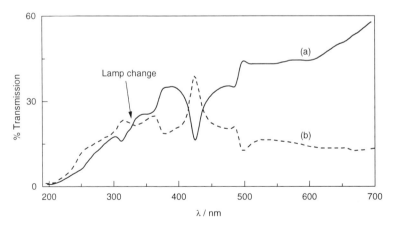

Fig. 5.8 Instrument polarization in a UV–visible spectrometer. UV–visible spectra of a dichroic film polarizer on a quartz substrate recorded on a Shimadzu UV250 double monochromator instrument: (a) plane of polarization horizontal; (b) plane of polarization vertical.

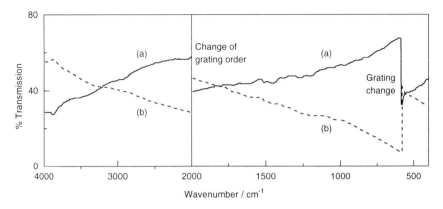

Fig. 5.9 Instrument polarization in a grating IR spectrometer. IR spectra of an aluminium grid polarizer on a KRS-5 substrate recorded on a Perkin-Elmer 684 spectrometer: (a) plane of polarization vertical; (b) plane of polarization horizontal. There is a change of grating order at 2000 cm^{-1} and a change of grating at 600 cm^{-1}.

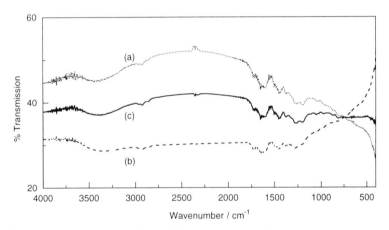

Fig. 5.10 Instrument polarization in an FTIR spectrometer. IR spectra of an aluminium grid polarizer on a KRS-5 substrate recorded on a Nicolet Impact 400D spectrometer: (a) plane of polarization vertical; (b) plane of polarization horizontal; (c) plane of polarization at 45° to the vertical. A spectrum recorded with the polarizer at $-45°$ ($= 135°$) to the vertical (not shown) was virtually identical to (c).

2.2.2 Choice of laboratory axes

Adopting the convention that the beam direction for photolysis or spectroscopy defines the laboratory Z-axis (see Section 1.6), it would be natural to adopt vertical and horizontal directions for the laboratory X- and Y-axes. In view of the significant beam polarization of spectrometers, however, this would be a bad choice. Instead, the conventional laboratory X- and Y-axes should be chosen at 45° to the vertical (Fig. 5.11). This choice minimizes the

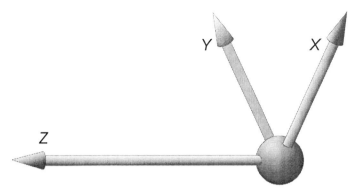

Fig. 5.11 Recommended laboratory axes for matrix experiments with plane-polarized light. The horizontal *Z*-axis represents the beam direction for matrix photolysis or spectroscopy. Note that the *X*- and *Y*-axes, at 45° to the vertical, are chosen to comply with the convention for right-handed axes (see Fig. 5.5).

differences in transmission between the *X*- and *Y*-polarizations, or, in other words, between the *X*-polarized and *Y*-polarized background spectra for each experiment. This can be seen for an IR spectrometer in Fig. 5.10 and for a UV–visible spectrometer in a comparison of Fig. 5.7(c) with Fig. 5.8.

2.3 Conducting matrix experiments with polarized light

Once the necessary polarizers have been obtained, carrying out matrix photolysis and spectroscopy with plane-polarized light is straightforward. Nevertheless, since polarizers cut out at least 50% of the light falling on the matrix, and usually a great deal more, photolysis times in polarization experiments tend to be long. This is especially the case when a monochromator or narrow-band interference filter is used, which cuts down the incident light even further. It is all too easy to photolyse with the 'wrong' polarization after hours of photoselection with the correct polarization, and thus wipe out an experiment. There is a need, therefore, for meticulous attention to detail at all times.

In the author's laboratory, a model set of axes like those shown in Fig. 5.11 was constructed from a wooden ball and three wooden rods about 40–50 cm long. The axes were labelled clearly as *X*, *Y*, and *Z*, and the whole model was suspended from the ceiling on nylon lines, close to the matrix-isolation equipment. Although this may smack of the nursery or play-school, and will no doubt attract occasional, uninformed ridicule, such a physical model provides a constant visual reminder of the directions for *X* and *Y* polarizations.

It is convenient if all spectrometer beams and the photolysis beam travel in the same direction. When viewed from the front, many spectrometers have a right-to-left beam direction, and the photolysis beam can be set up in a right-to-left direction as well. Not all spectrometers have this optical arrangement, however. If polarization experiments are being conducted with both right-to-left and left-to-right beams, the cold window can obviously be rotated through 180°, so that each beam passes through the matrix in the same direction. There is no need for this rotation, however, and it is simpler, and probably less confusing, to allow beams to pass through the matrix in either direction, while keeping the *XZ*- and *YZ*-polarization planes fixed absolutely within the laboratory frame. In other words some beams will travel in the *Z*- and others in the $-Z$ direction.

Finally, it should be noted that grating monochromators are polarized in the same way as spectrometers. If a grating monochromator is to be used in polarization experiments, maximum light intensity will only be achieved with the monochromator at the correct angle. With the laboratory *X*- and *Y*-axes at 45° to the vertical, this will usually mean having the base of the monochromator at 45° to the vertical also. Moreover, the monochromator will need to be rotated along with the polarizer, if a change in direction of polarization of the photolysis beam is needed for any reason.

2.3.1 Standard procedures for matrix experiments with polarized light

In some specialized studies it may be necessary to record a series of polarized spectra with the planes of polarization at closely spaced angles, for example every 5°. In most experiments, however, it is sufficient to record just the *X*- and *Y*-polarized spectra, although some workers also like to record an unpolarized spectrum at each stage, as a check. Protocol 1 gives an outline of the procedure for a generalized matrix photolysis experiment using plane-polarized light.

Protocol 1.
Matrix photolysis and spectroscopy with plane-polarized light

The following outline procedure is appropriate for polarization experiments with either IR or UV–visible spectrometers.

Equipment
- A matrix-isolation cold cell
- A UV–visible polarizer for the photolysis source
- An appropriate spectrometer (IR or UV–visible)
- A polarizer for the spectrometer
- A reasonably well collimated light source, such as a high-pressure Hg arc with a quartz lens, together with filters or a monochromator, if necessary

Materials
- A suitable matrix-gas mixture with a photolabile guest, or research grade host gas and a separate guest material

1. Set up the cold cell and spray-on line and cool the cell to the temperature at which matrix deposition is to be carried out (usually the base temperature).

2. If an unpolarized spectrum is to be recorded, locate the cold cell in the sample compartment of the spectrometer and align the cold window optimally with the beam. Record the unpolarized background spectrum and save it as a reference spectrum.

3. Mount the spectrometer polarizer in the sample beam (or on the shroud of the cold cell, if this position is preferred). This may require temporary removal of the cold cell from the sample compartment.

4. Record background spectra in both X and Y polarizations and save each as a reference spectrum. If possible, adjust the polarizer between spectra without moving the cold cell. This will avoid spurious differences between the X- and Y-polarized spectra due to inconsistent positioning of the cold window.

5. Subtract the appropriate background spectrum (X polarized, Y polarized or unpolarized) from each subsequent spectrum recorded in the experiment. This can be done for all the spectra at the end of the experiment, or by using the appropriate background as a reference while each spectrum is recorded.

6. Once the background spectra have been recorded and saved, spray-on the matrix as described in Chapter 4, for example following Protocol 7.

7. When a sufficiently thick matrix has been deposited, record the X- and Y-polarized spectra (and an unpolarized spectrum if desired). At this stage, there should be no significant difference between the X and Y spectra (or the unpolarized spectrum) after subtraction of the appropriate backgrounds; if there seems to be a difference, run the spectra again, making sure that the correct background is used as a reference in each case.

8. Set up the light source with the best possible beam collimation, and with filters or a monochromator in place, if these are to be used.

9. Position the UV–visible polarizer in the photolysis beam and adjust it to give either X- or Y-polarization. In some experiments, X and Y polarized photolysis may be carried out in sequence. Usually, however, photolysis will be carried out using just one polarization throughout. In a series of polarization experiments, it is a good idea to vary the selected photolysis polarization routinely between the X and Y directions, that is, not to use X polarization, say, every time. This provides a check for any unwanted polarization effects, such as might arise from optical components in the photolysis beam, for example.

10. Expose the matrix to the polarized light for an appropriate time interval.

11. Record X- and Y- polarized spectra.

12. Continue the sequence of polarized photolysis and the recording of polarized spectra for as long as necessary. Experiments of this type can last for several days.

13. Finish by evaporating the matrix and warming the cold cell in the usual way (see Chapter 4, Protocol 7 and Section 3.2).

2.4 Examples of polarized matrix IR spectroscopy

In the previous sections of this chapter, some examples of polarized matrix UV–visible spectra have been given (Figs 5.3 and 5.4). One of the main advantages of matrix photoselection, however, is derived when IR absorptions are studied in the same way. The following two examples have been selected to illustrate this.

2.4.1 Diazocyclopentadiene

When diazocyclopentadiene (CpN$_2$, Scheme 5.1) is photolysed in a matrix, the reactive carbene, cyclopentadienylidene (Cp:), is generated. In CO, which is a reactive matrix host, the carbene reacts immediately with one of the surrounding CO molecules, to give the ketene, CpCO. In an inert or nearly inert matrix such as Ar or N$_2$, however, the carbene persists, and its UV and IR spectra can be recorded.[10] Figure 5.12 shows polarized IR bands of CpN$_2$ and the carbene, observed after partial polarized photolysis of CpN$_2$ in an N$_2$ matrix. The IR absorptions shown belong to the very strong diazo stretch (v(CNN)) of CpN$_2$ and the out-of-plane C–H deformations (γ(CH) vibrations) of CpN$_2$ and the carbene. From Fig. 5.6(a) it will be apparent that the v(CNN) band of CpN$_2$ has a transition moment in the molecular z-direction, and is

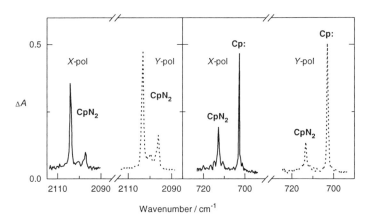

Fig. 5.12 IR linear dichroism in the matrix photolysis of diazocyclopentadiene. Diazo-cyclopentadiene (Scheme 5.1, CpN$_2$) in an N$_2$ matrix (N$_2$:CpN$_2$ = 5000:1) at 12 K was partially photolysed with X-polarized light (λ = 315 ± 10 nm). The spectra show the split v(C=N=N) band of CpN$_2$ near 2100 cm^{-1} and the γ(C-H) (out-of-plane) bands of CpN$_2$ (713 cm^{-1}) and the carbene product, Cp: (703 cm^{-1}), recorded with X- (solid lines) and Y- (dotted lines) polarization. Note that: (a) all three components of the split v(C=N=N) band of CpN$_2$ are parallel polarized ($A_X < A_Y$), (b) the γ(C–H) band of CpN$_2$ is perpendicularly polarized ($A_X > A_Y$), and (c) the γ(C–H) band of Cp: has a lower degree of polarization and the opposite sense of polarization compared with the γ(C–H) band of CpN$_2$, as expected for a photoproduct which has undergone partial rotational randomization. *(Adapted with permission from ref. 10.* © *1981, American Chemical Society.)*

therefore of A_1 symmetry; while the out-of-plane C-H deformations are of B_1 symmetry with transition moments in the molecular x-direction.

In Fig. 5.12 it can be seen that the residual, photoselected CpN_2 showed a fair degree of polarization. After photolysis with X-polarized light, the A_1 $\nu(CNN)$ band of CpN_2 clearly exhibited parallel polarization ($A_X < A_Y$), from which it could be concluded that the photo-active electronic transition of CpN_2 also has A_1 symmetry, i.e. with its transition moment in the molecular z-direction. On the other hand, the $\gamma(C–H)$ band of CpN_2 exhibited perpendicular polarization ($A_X > A_Y$), as expected for a vibrational mode of B_1 symmetry.

With a C_{2v} molecule such as CpN_2, the observed linear dichroism in the IR spectrum will divide the vibrational bands into two groups: those with parallel polarization and those with perpendicular polarization. In the case of CpN_2, the first group belong to vibrational modes with A_1 symmetry, and the second group to modes with either B_1 or B_2 symmetry. The observed symmetry of the parallel polarized bands defines the symmetry of the photo-active electronic transition, and this in turn allows some deductions to be made concerning the electronic nature of the excited state. Simple molecular orbital calculations for CpN_2 suggested that the only low-energy electronic transition which has A_1 symmetry involved promotion of an electron from the doubly occupied $3b_1$ orbital into the unoccupied $4b_1$ orbital (Fig. 5.13); the transition moment for this transition has A_1, symmetry because both orbitals have the same symmetry. Note that the electron is promoted from an orbital which is bonding between C–1 and N to one which is anti-bonding, and this fits in well with the C–N cleavage which occurs upon excitation. It should, however, be re-membered that the experimentally observed UV absorption is a vibronic envelope, and does not correspond to a purely electronic transition. In general, the observed symmetry of a vibronic absorption will depend on the symmetries of both the electronic and vibrational transitions involved. Some caution is therefore needed when interpreting linear dichroism observations in terms of molecular orbitals. Nevertheless, photoselection experiments do provide a means of testing molecular orbital calculations experimentally.

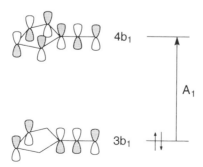

Fig. 5.13 A molecular orbital representation of the photoactive electronic transition of A_1 symmetry in diazocyclopentadiene (CpN_2).

Ian R. Dunkin

Finally, Fig. 5.12 also reveals that the carbene generated by photolysis of CpN$_2$ was photo-oriented to a certain extent. The γ(CH) band belonging to the carbene exhibited distinct parallel polarization ($A_X < A_Y$ following X-polarized photolysis). This is the sense of polarization expected for an out-of-plane (B$_1$) mode of a photo-oriented product. The degree of polarization of this Cp: band appears to be lower than that observed for the CpN$_2$ bands, and this may indicate that a certain amount of molecular rotation, and thus directional randomization, occurred after photolysis. A similar observation was made in the UV spectra of photo-oriented CpCO molecules (see Section 1.3 and Fig. 5.3).

2.4.2 Pentacarbonylchromium

One of the first studies of polarized photolysis in matrices, and certainly the first to be published, was the photo-orientation and photo-reorientation of pentacarbonylchromium.[23] This coordinatively unsaturated species can be generated by the photolysis of Cr(CO)$_6$ and has C$_{4v}$ symmetry (square pyramidal). As described in Chapter 1 (see Section 2.2.4 and Table 1.1), Cr(CO)$_5$ forms complexes with a range of matrix host molecules, and its lowest energy electronic transition is very sensitive to this complexation, λ_{max} varying from 624 nm in Ne matrices to 489 nm in CH$_4$ matrices. Irradiation into the visible band of such a Cr(CO)$_5$ complex with CH$_4$ results in partial reversal to Cr(CO)$_6$; this provided a way to generate an oriented sample of Cr(CO)$_5$, by photoselection with plane-polarized light (Scheme 5.3).

$$Cr(CO)_6 \underset{h\nu \text{ - visible}}{\overset{\substack{h\nu \text{ - UV} \\ CH_4 \text{ matrix}}}{\rightleftharpoons}} CH_4 \cdots Cr(CO)_5 + CO$$

Scheme 5.3

Figure 5.14(a) shows the results of such photoselection experiments with Cr(CO)$_5$ in a CH$_4$ matrix. The visible electronic absorption at 489 nm exhibited the expected parallel polarization ($A_Y < A_X$ after Y-polarized photolysis), while the ν(CO) IR band at 1932 cm^{-1} exhibited perpendicular polarization. When the matrix was irradiated further with X-polarized light, photo-reorientation of the Cr(CO)$_5$ molecules was observed (Fig. 5.14(b)).

In C$_{4v}$ molecules such as Cr(CO)$_5$, the allowed transitions have either A$_1$ (z) or E (xy) symmetry. The latter modes are degenerate, and for these, the transition moments do not have a single direction within the molecular framework, but have equal probability of photon absorption in all directions lying in the molecular xy-plane. The IR band shown in Fig. 5.14 is known to belong to an axial CO stretch, with its transition moment in the molecular z-direction, and therefore to be of A$_1$ symmetry. Since this band exhibits perpendicular polarization, it could be concluded that the photoactive electronic absorption has E symmetry, which agreed with a previous of

164

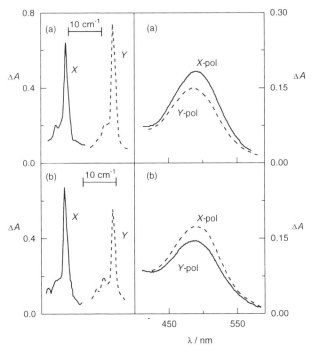

Fig. 5.14 Photo-orientation and photo-reorientation of $Cr(CO)_5$. Left: X- and Y-polarized IR absorptions of the A_1 (axial) $\nu(C\text{-}O)$ band of $Cr(CO)_5$ at 1932 cm^{-1}. Right: X- and Y-polarized visible absorptions of $Cr(CO)_5$ with λ_{max} = 489 nm. The spectra were recorded after unpolarized UV photolysis of $Cr(CO)_6$ in a CH_4 matrix (CH_4:$Cr(CO)_6$ = 5000:1) at 20 K to generate the $Cr(CO)_5$, followed by two periods of polarized visible photolysis ($\lambda > 375$ nm): (a) for 15 minutes with Y-polarized light; (b) for a further 15 minutes with X-polarized light. *(Adapted with permission from ref. 23. © 1975, Royal Society of Chemistry.)*

assignment of this transition as $^1A_1(b_2^2e^4)\rightarrow{}^1E(b_2^2e^3a_1^1)$, that is, a transition with a transition moment of E symmetry.

In Fig. 5.15, pentacarbonylchromium is shown simply as a square representing the xy plane of the molecule, with a perpendicular line representing the molecular z-axis. Three extreme orientations of the molecule are shown: with the molecular z-axis aligned with the laboratory X-, Y-, and Z-axes. With X-polarized light, Y- and Z-oriented $Cr(CO)_5$ molecules can absorb, because the E-vector of the light lies in the molecular xy-plane in each case, but the X-oriented molecules have the molecular xy-plane perpendicular to the E-vector and therefore cannot absorb. With Y-polarized light, the X- and Z-oriented molecules can absorb, but not the Y-oriented molecules.

Photo-reorientation of $Cr(CO)_5$ probably occurred without dissociation into $Cr(CO)_4$ and CO. Calculations predicted a D_{3h} structure for the excited state of $Cr(CO)_5$ and this D_{3h} structure could decay back to the C_{4v} ground

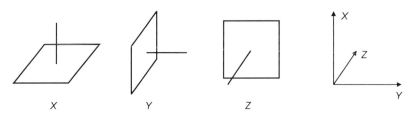

Fig. 5.15 Orientations of Cr(CO)$_5$. The Cr(CO)$_5$ molecule is shown as a square representing the molecular xy-plane (with a CO ligand at each corner) and a line indicating the unique axial CO lying in the molecular z-direction. The three orientations shown have the molecular z-axis aligned with the laboratory X-, Y-, and Z-axes. The photoactive electronic transition of Cr(CO)$_5$ has E symmetry, which means that there is equal probability of photon absorption with the **E**-vector of the incident light at any angle in the molecular xy-plane.

state in several different orientations, thus resulting in an overall molecular rotation (Scheme 5.4).

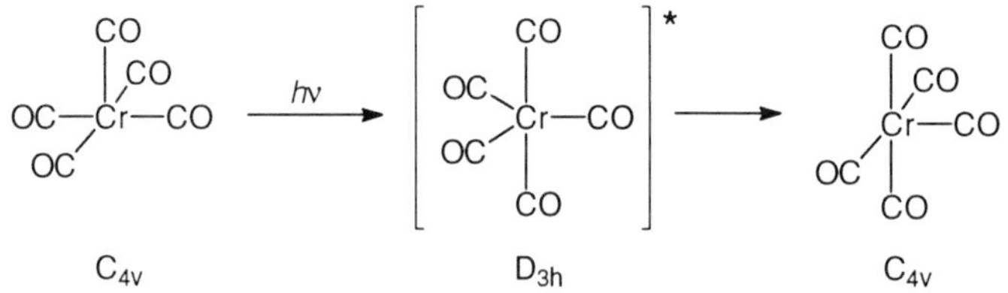

Scheme 5.4

2.5 The interpretation of IR dichroism in molecules of low symmetry

The few examples discussed so far in this chapter have all involved molecules of at least C$_{2v}$ symmetry. In these cases, interpretation of dichroic spectra is greatly facilitated because the transition moments—both electronic and vibrational—are restricted by symmetry considerations to a small number of directions within the molecules. For example, in C$_{2v}$ molecules all transition moments lie along one of three mutually orthogonal molecular axes: x, y, or z. With a partially oriented array of C$_{2v}$ molecules, all absorptions will show linear dichroism, either parallel or perpendicular. Moreover, all the parallel polarized bands should have the same degree of polarization, within experimental error. Similarly, all the perpendicularly polarized bands should have a consistent degree of polarization, which is slightly lower than that for the parallel bands. One consequence of this is that it is generally fairly easy to determine whether a molecule has lower symmetry than C$_{2v}$.

Polarized photolysis of matrix-isolated Mn$_2$(CO)$_{10}$, for example, produced partially oriented Mn$_2$(CO)$_9$ which, from the IR spectrum, clearly possessed a bridging carbonyl group.[24] The terminal ν(CO) bands of the oriented

$Mn_2(CO)_9$ exhibited distinct linear dichroism, but the $\nu(CO)$ band of the bridging carbonyl was unpolarized within experimental error. As explained above, this pattern of dichroism is inconsistent with a C_{2v} structure, which led to the conclusion that $Mn_2(CO)_9$ has a C_s structure with an η^1,η^2 semi-bridging CO group rather than a C_{2v} structure with a normal bridging carbonyl (Scheme 5.5).

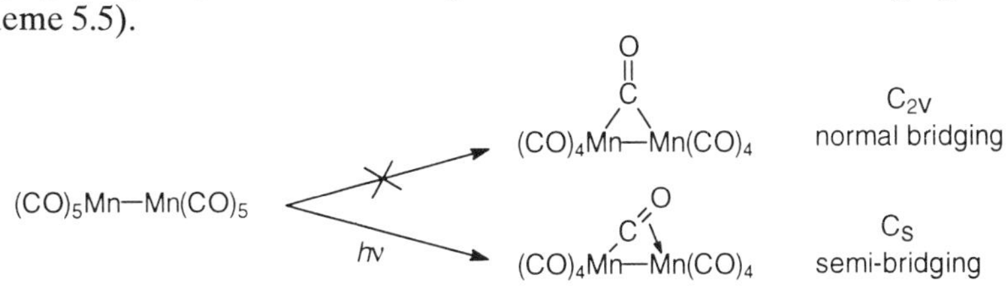

$$(CO)_5Mn—Mn(CO)_5 \xrightarrow{h\nu}$$

$$\begin{array}{c} O \\ \| \\ C \\ / \ \backslash \\ (CO)_4Mn—Mn(CO)_4 \end{array} \quad \begin{array}{c} C_{2v} \\ \text{normal bridging} \end{array}$$

$$(CO)_4Mn—Mn(CO)_4 \quad \begin{array}{c} C_s \\ \text{semi-bridging} \end{array}$$

Scheme 5.5

More recently, Josef Michl and co-workers have studied the minor conformer of buta-1,3-diene.[25] It is well known that the most stable butadiene conformer is the planar *s-trans* form (Fig. 5.16(a)), but the structure of the minor conformer—present at only about 3% in the equilibrium mixture at room temperature—has been more difficult to determine. Matrix isolation of butadiene which had been passed through an oven at 1200 K, followed by photoselection with 248 nm light, gave a partially oriented sample of the minor conformer. The observed IR dichroism provided convincing evidence that this conformer has the planar C_{2v} *s-cis* structure (Fig. 5.16b), in disagreement with theoretical predictions, which tended to favour a *gauche* C_2 structure with a $C_1C_2C_3C_4$ dihedral angle of about 40° (Fig. 5.16(c)).

2.5.1 Qualitative interpretation of linear dichroism for molecules of low symmetry

If an array of low-symmetry molecules is partially oriented by photoselection, photo-orientation or photo-reorientation, the electronic and vibrational absorptions of the sample will, in general, exhibit linear dichroism. The degree of polarization of each absorption band will, however, vary according to the direction of its transition moment with respect to that of the photo-active transition.

It is conventional to take the direction of the photoactive absorption as defining the molecular z-axis. The highest degree of polarization will then be found for absorptions with transition moments lying at small angles to the

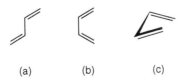

(a)   (b)   (c)

Fig. 5.16 Conformers of buta-1,3-diene. (a) *s-trans*; (b) *s-cis*; (c) *gauche*.

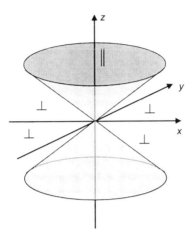

Fig. 5.17 Molecular x-, y-, and z-axes and directions in which transition moments give rise to parallel (∥) and perpendicularly (⊥) polarized absorptions. It is assumed that photoselection has been achieved by photolysis into an absorption with its transition moment in the z direction. The double cone has a vertical angle of 108°, and defines the direction of transition moments (54° with respect to the z-axis) corresponding to absorptions with zero dichroism. *(Reproduced with permission from ref. 21. © 1997, Elsevier.)*

z-axis. These absorptions will be polarized in the same sense as the photo-active absorption, that is, with $A_Y > A_X$ after X-polarized photolysis, and are referred to as parallel polarized absorptions. As the angle of the transition moment with respect to the molecular z-axis increases, the degree of polarization of the corresponding absorption decreases, becoming zero when the angle reaches 54°. At angles greater than 54°, the degree of polarization increases again but the polarization is now perpendicular rather than parallel, that is, with $A_Y < A_X$ after X-polarized photolysis. The highest degree of perpendicular polarization will be observed for absorptions with transition moments lying in the molecular xy-plane. Figure 5.17 shows the directions of transition moments within the molecular framework which give rise to parallel and perpendicular polarizations. The two groups of transition moments are separated by a double cone with a vertical angle of 108°. Transition moments with directions lying anywhere on the surface of the double cone will correspond to absorptions with no dichroism. Transition moments lying close to the surface of this cone will be only weakly dichroic.

Careful inspection of Fig. 5.17 will reveal some pitfalls of interpretation that must be avoided:

(a) Two transition moments can be at 90° to each other and still lie within the double cone. Although perpendicular to each other, they will both give parallel polarized absorptions, though the degree of polarization will be small.

(b) Two transition moments can be at 90° to each other and both be perpendicularly polarized; for example, any orthogonal pair in the xy plane.

(c) Two transition moments can lie close to each other, one just inside the cone and the other just outside, resulting in one having parallel and the other perpendicular polarization. Of course, the degree of polarization will be small if the transition moments are very close to the surface of the cone.

These points emphasize that the observed dichroism is related to the photoactive transition, and it cannot be concluded that any two transition moments with the same direction of polarization are approximately collinear or that any two transition moments with opposite polarization are approximately orthogonal.

The relationship between the extent of molecular orientation achieved in experiments with polarized photolysis and the resulting linear dichroism can be treated quantitatively,[13,18] but in practice the experimental measurement of degrees of polarization for matrix-isolated samples suffers from some inaccuracies which are difficult to quantify. These originate from such factors as the precise choice of baseline, overlap between bands, and the possible partial depolarization of IR or UV–visible beams due to scattering. It has therefore been the author's practice to rely as much as possible on qualitative deductions from dichroic spectra.

In purely qualitative terms the absorptions of the photolytically oriented sample will fall into three classes:

- those with parallel polarized bands
- those with perpendicularly polarized bands
- those showing no dichroism or only very weak dichroism.

For molecules with C_{2v} symmetry, all transition moments for allowed transitions must lie along the molecular x-, y-, or z-axes, so there will be no absorptions in the third class. Indeed, as has been shown above for the case of $Mn_2(CO)_9$, the existence of one or more bands with zero or near zero degree of polarization while other bands are significantly dichroic, demonstrates that a molecule does not have C_{2v} symmetry.

For molecules of lower symmetry or no symmetry at all, a graphical mode of interpretation for the linear dichroism of IR bands may be helpful,[21,22] as will now be explained. For some IR bands, for example the stretching modes of cyanide or carbonyl groups, the directions of the transition moments within a molecule may be fairly obvious, at least to a good approximation. A cyanide group, for example, is likely to have the transition moment of its $\nu(CN)$ band collinear with the C≡N bond, to a very close approximation. Therefore the observed linear dichroism of each of these bands gives an estimate of the direction of the transition moment of the photoactive transition. Allowing a generous margin for uncertainties, one can conclude the following.

(a) Bands which are distinctly parallel polarized will have transition moments which lie within 45° of the photoactive transition moment.

(b) Bands with small degrees of polarization will have transition moments between, say, 35° and 75° of the photoactive transition moment.

(c) Bands which are distinctly perpendicularly polarized will have transition moments at more than 60° to that of the photoactive transition.

These rather crude estimates can be refined if the transition moments of individual vibrational modes can be determined with greater accuracy, such as by normal mode analysis. In any case, each observation of linear dichroism in the IR bands of a photoselected, photo-oriented, or photo-reoriented sample yields a possible range of directions for the transition moment of the photo-active absorption. The various estimations of the direction of the photo-active transition moment should, of course, be consistent within the accepted margin of error. If they are not, then the observed linear dichroism is not consistent with the proposed structure or conformation of the molecule being studied.

An illustration of the application of this approach is given by a recent study of the conformations of matrix isolated cyano-substituted phenyl azides.[21,22] Phenyl azides are known to adopt planar or nearly planar conformations, but an unsymmetrically substituted phenyl azide has two different planar conformers, which can be interconverted by rotation about the C–N bond. 3-Cyano-2-methylphenyl azide, for example, can exist as the planar *syn* and *anti* forms shown in Fig. 5.18. On steric grounds, the *anti* conformer would be expected to be the more stable and therefore to predominate, but experimental evidence which can identify the predominant conformer in a low-temperature matrix is not so easy to come by.

Photoselection of 3-cyano-2-methylphenyl azide in an N_2 matrix produced a dichroic sample in which the $v(CN)$ band of the cyanide group was perpendicularly polarized, the pseudosymmetric azide stretch ($v(NNN)_{ps}$) was parallel polarized, and the asymmetric azide stretch ($v(NNN)_{as}$) was not discernibly polarized at all.[22] Assuming in the first instance that the matrix-isolated azide adopts the expected *anti* conformation, Fig. 5.19 shows how each of these observations of IR linear dichroism leads to a prediction of the range of possible directions for the photo-active electronic transition. All three predictions are reasonably consistent. Figure 5.20 shows a similar analysis of the dichroism assuming that the molecule adopts the *syn* conformation. In this

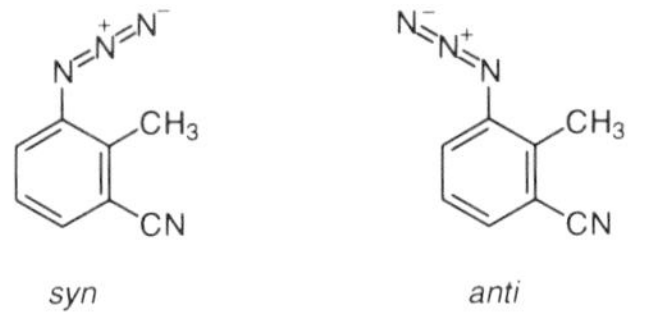

Fig. 5.18 The two planar conformers of 3-cyano-2-methylphenyl azide.

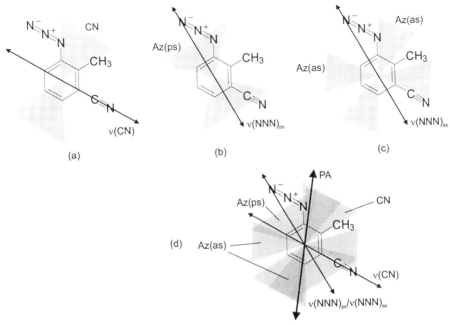

Fig. 5.19 Estimates of the transition moment direction of the photo-active transition for the *anti* conformer of 3-cyano-2-methylphenyl azide. In (a), (b), and (c), best estimates of the vibrational transition moment directions for $v(CN)$, $v(NNN)_{ps}$, and $v(NNN)_{as}$, respectively, are shown as double-headed arrows. The probable limits for the direction of the transition moment of the photoactive transition, derived from the observed dichroism in each case, are shown as shaded areas: CN (perpendicular polarization), Az(ps) (parallel polarization), and Az(as) (zero polarization), respectively. In (d), all three estimates for the transition moment of the photoactive transition are superimposed, showing acceptable overlap. The overall best estimate for the direction of the photoactive transition is shown as the double headed arrow, PA.

latter case, however, the predictions for the direction of the photoactive transition are not consistent. It can therefore be concluded that the preferred conformation or matrix-isolated 3-cyano-2-methylphenyl is the expected *anti* form.

Note that in Figs 5.19 and 5.20 the estimated transition moments for the azide pseudosymmetric and asymmetric stretches are shown as collinear, at about 30° to the C–N bond between the ring and the azide group. At present, this is the best estimate available, but since the IR absorptions of these two vibrational modes do not have the same dichroism, it is clearly a gross approximation. Reliable computations of transition moment directions for vibrational modes would considerably improve the estimates for the direction of the photoactive transition.

Qualitative analyses of patterns of dichroism for molecules of low symmetry

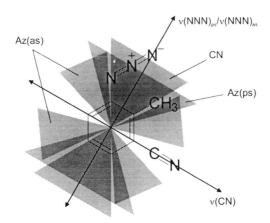

Fig. 5.20 Estimates of the transition moment direction of the photoactive transition for the *syn* conformer of 3-cyano-2-methylphenyl azide. The diagram shows the superimposition of estimates of the probable limits for the transition moment direction of the photoactive transition, derived from the observed dichroism of the $\nu(CN)$, $\nu(NNN)_{ps}$ and $\nu(NNN)_{as}$ bands. The probable limits in each case are shown as shaded areas: CN, Az(ps), and Az(as), respectively (cf. Fig. 5.19). Note that there is no overlap between the shaded areas corresponding to the $\nu(CN)$ and $\nu(NNN)_{as}$ bands. The observed dichroism is therefore inconsistent with the *syn* conformation.

will not always be so clear cut as in the example just discussed, and in any case will seldom lead to completely rigorous conclusions about molecular structures. In more difficult cases, detailed computations of the transition moment directions for various vibrational modes would improve the discrimination of the technique.

2.6 Conclusion

From a perusal of this chapter, the reader may well have decided that in matrix studies with polarized light, carrying out the experiments is more straightforward than interpreting the results. In many cases this could be true. Perhaps it is one of the main reasons why polarized matrix photolysis has not yet been exploited very widely. It is certainly the reason why the author has spent so much space in this chapter on problems of theory and interpretation as well as the more practical aspects of the subject.

References

1. Weigert, F. *Ber. Dtsch Physik. Ges.* **1919**, *21*, 479–491; 615–622; 623–631.
2. Weigert, F. *Z. Physik.* **1920**, *3*, 437–459; **1921**, *5*, 410–427.
3. Weigert, F. *Z. Physik. Chem., Abt. B.* **1929**, *3*, 377–404.
4. Weigert, F. *Z. Electrochem.* **1921**, *27*, 481–487.
5. Weigert, F.; Nakashima, M. *Z. Physik. Chem., Abt. B* **1930**, *7*, 25–469.

6. Lewis, G. N. and Bigeleisen, J. *J. Am. Chem. Soc.* **1943**, *65*, 520–526.
7. Albrecht, A. C. *J. Chem. Phys.* **1957**, *27*, 1413–1414.
8. Albrecht, A. C. *J. Mol. Spectrosc.* **1961**, *6*, 84–108.
9. Albrecht, A. C. *Progr. React. Kinet.* **1970**, *5*, 301–334.
10. Baird, M. S.; Dunkin, I. R.; Hacker, N.; Poliakoff, M.; Turner, J. J. *J. Am. Chem. Soc.* **1981**, *103*, 5190–5195.
11. Turner, J. J.; Burdett, J. K.; Perutz, R. N.; Poliakoff, M. *Pure Appl. Chem.* **1977**, *49*, 271–285.
12. Bauer, G. *Monatsh. Chem.* **1971**, *102*, 1782–1788; 1789–1796.
13. Michl, J.; Thulstrup, E. W. *J. Chem. Phys.* **1980**, *72*, 3999–4008; *Acc. Chem. Res.* **1987**, *20*, 192–199; **1988**, *21*, 94.
14. Thulstrup, E. W.; Michl J. *J. Mol. Struct.* **1980**, *61*, 175–182.
15. Friesner, R.; Nairn, J. A.; Sauer, K. *J. Chem. Phys.* **1980**, *72*, 221–230.
16. Nairn, J. A.; Friesner, R.; Sauer, K. *J. Chem. Phys.* **1981**, *74*, 5398–5406.
17. Radziszewski, J. G.; Burkhalter, F. A.; Michl, J. *J. Am. Chem. Soc.* **1987**, *109*, 61–65.
18. Michl, J.; Thulstrup, E. W. *Spectroscopy with Polarized Light*; VCH: Weinheim, **1986**.
19. See, for example, Salthouse, J. A.; Ware, M. J. *Point Group Character Tables and Related Data*; Cambridge University Press, **1972**.
20. See, for example, Harris, D. C.; Bertolucci, M. D. *Symmetry and Spectroscopy*; Oxford University Press: New York, **1978**.
21. Dunkin, I. R.; Shields, C. J. *Spectrochim. Acta, Part A* **1997**, *53*, 129–140.
22. Dunkin, I. R.; Shields, C. J. *Spectrochim. Acta, Part A* **1997**, *53*, 141–150.
23. Burdett, J. K.; Perutz, R. N.; Poliakoff, M.; Turner, J. J. *J. Chem. Soc., Chem. Commun.* **1975**, 157–159.
24. Dunkin, I. R.; Härter, P.; Shields, C. J. *J. Am. Chem. Soc.* **1984**, *106*, 7248–7249.
25. Fisher, J. J.; Michl, J. *J. Am. Chem. Soc.* **1987**, *109*, 1056–1059; Arnold, B. R.; Balaji, V.; Downing, J. W.; Radziszewski, J. G.; Fisher, J. J.; Michl, J. *J. Am. Chem. Soc.* **1991**, *113*, 2910–2919.

173

6

Classic matrix experiments

Anyone embarking on matrix-isolation studies will want to get on with their own ideas and experiments as soon as possible. There will be little incentive to spend time repeating the work of others. Nevertheless, it is valuable to gain confidence in new equipment or new techniques with some proven examples; so this chapter brings together a selection of matrix experiments from the literature, which can be carried out as tests or simply for fun. The author has carried out a good proportion of the experiments, but by no means all.

The experiments have been chosen to illustrate as wide a range of matrix-isolation studies as possible, but without the requirement of very specialized equipment or long syntheses of starting materials. They are arranged in two main groups: (*i*) stable molecules and molecular complexes and (*ii*) reactive species. Within these groups, some of the experiments are described in a series of protocols, each of which is really an outline procedure which refers to detailed advice and instructions given in previous chapters. At the beginning of each protocol will be found the usual list of necessary equipment and materials, and also lists of any specialized techniques involved and aspects of matrix-isolation research which the experiment illustrates. In other cases, the proposed experiments are little more than suggestions, with pointers to the original literature.

All the experiments require a cold cell, as described in Chapter 2, and the means of making up gas bulbs and depositing matrices (see Chapters 3 and 4); to avoid needless repetition, these items are omitted from the equipment lists unless a special feature is needed. It is assumed that the cold cell is fitted with a CsBr or CsI cold window and KBr external windows, unless stated otherwise.

Most of the experiments could be carried out without reference to the original papers. In keeping with the aims of this book, however, background information has been kept to a minimum, and the emphasis in this chapter, as elsewhere, is on practical techniques and the testing of equipment. The identification of guests trapped in matrices, be they complexes formed from stable molecules or highly reactive species, usually requires a great deal of painstaking research, often involving isotopic substitution or detailed computations. For a complete picture, therefore, the reader is encouraged to consult at least the key papers relating to a particular experiment.

1. Stable molecules and molecular complexes in matrices

So far as equipment is concerned, studies of stable molecules in matrices require only a cold cell, gas-handling facilities, and a suitable spectrometer. There is no need for photolysis sources, pyrolysis tubes, or other methods of generating reactive species. Therefore, experiments with stable molecules, or complexes formed from stable molecules, can be an attractive way of testing a new matrix-isolation set-up, especially if the starting materials are readily and cheaply available. There is one caveat, however. The interpretation of matrix IR spectra of molecules which can readily form H-bonds can be plagued by the presence of water in the matrices. For instance, in annealing experiments, where diffusion of the trapped species takes place, it can be difficult to ascertain whether bands which arise in this process belong to dimers or multimers of the intended guest or to complexes between this guest and water present as an impurity. In studies of this type, therefore, it is essential to reduce the presence of background water vapour in the vacuum systems to the minimum level possible.

1.1 Stable molecules in matrices

1.1.1 Hydrogen chloride in Ar matrices

Matrix-isolated hydrogen chloride was extensively investigated in the 1960s. The HCl molecule is very simple—it has only one fundamental vibration—but its matrix IR spectra have a number of interesting features, which have been discussed in detail by Hallam.[1] The suggested experiment is to prepare a mixture of Ar and HCl with a matrix ratio (Ar:HCl) in the range 50:1 to 2000:1, deposit this as a matrix at 20 K, and record the IR spectra before and after annealing. The overall procedure for the experiment is given in Protocol 1 below, and was first described by Barnes *et al.*[2] The results which can be expected are discussed immediately following the protocol.

Protocol 1.
IR spectra of HCl in Ar matrices

Caution! HCl is both an irritant and corrosive, and should not be allowed to escape from the cylinder into the open laboratory. Fit the correct type of stainless steel valve to the cylinder, store the cylinder in a fume cupboard or other well ventilated place when it is not in use, and vent any excess HCl into a fume cupboard or to the outside of the building. When making up the matrix-gas mixture, avoid exposing expensive and delicate gauge heads to the HCl; use an inexpensive Bourdon gauge or a manometer (cf. Fig. 3.2) instead.

Equipment

- Cold cell equipped with a heater and temperature controller
- IR spectrometer
- Calibrated expansion bulb (optional)

Materials

- HCl of at least 99% purity (cylinder)   **irritant, corrosive**
- High purity Ar

Illustrates

- Matrix IR spectroscopy
- Annealing of matrices
- Dimer and multimer formation

1. Set up a preparative vacuum line (cf. Figs 4.1 and 4.4) with a 1 litre gas bulb and cylinders of Ar and HCl. (See Chapter 4, Section 2.4.4 for advice on connecting the two cylinders.) Include a calibrated expansion bulb if the gas mixture to be prepared is to have a matrix ratio (Ar:HCl) of 500:1 or greater.

2. Prepare a gas mixture of Ar and HCl with Ar:HCl in the range 50:1 to 2000:1, following Protocols 3 or 6 in Chapter 4.

3. Transfer the bulb containing the gas mixture to the spray-on line of the cold cell.

4. Following Protocol 7 in Chapter 4, deposit the gas mixture as a matrix at 20 K by the slow deposition technique and record the IR spectrum once a sufficiently thick matrix has built up.

   NB The original paper gave 20 K as the temperature for deposition, but deposition at the base temperature of the cold cell (e.g. 10–12 K) should work equally well, although some differences in the observed IR absorptions may result.

5. Anneal the matrix briefly at 35 K as follows:

   (a) Set the temperature controller for a moderate power output, for example 1–5 W.

   (b) Adjust the temperature setting to a few K below the desired annealing temperature of 35 K (30–33 K, say) and switch on the controller if it is not already operating.

   (c) Allow the temperature to stabilize just below 35 K for a few seconds, then adjust the temperature setting to 35 K.

   NB The precaution of approaching the final temperature in two steps avoids overshooting, but may not be necessary with some controllers and heaters.

   (d) Keep the matrix at 35 K for a few minutes, then return the temperature setting to its original value (e.g. 20 K) and allow the matrix to cool and stabilize at the lower temperature.

6. Record the IR spectrum again.

Protocol 1. *Continued*

7. The experiment may be continued with further periods of annealing, recording an IR spectrum after each one. Both the duration and temperature of annealing may be increased, although matrix boil-off will become excessive above about 40 K.

8. The experiment may be repeated with different Ar:HCl ratios, if desired.

Anhydrous gaseous HCl of at least 99% purity can be purchased in conveniently sized cylinders. Alternatively, it can be generated by adding concentrated hydrochloric acid to concentrated sulphuric acid and dried by passage through phosphorus pentoxide; it is then necessary to purify it by trap-to-trap distillation on a vacuum line.

The sort of IR spectra that are expected for HCl in Ar matrices are shown in Fig. 6.1. The reader should consult the original paper and earlier work quoted therein for a full analysis.[2] At a matrix ratio (Ar:HCl) of 2000:1, the HCl is present almost completely as isolated molecules. The spectrum is that of the monomeric species, with a weak additional band at 2818 cm^{-1} due to a small

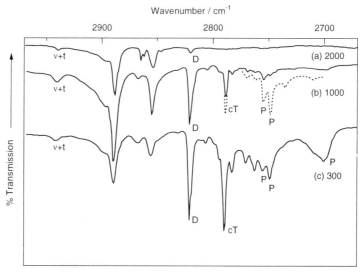

Fig. 6.1 IR spectra of the ν(H–Cl) region of HCl in Ar matrices at 20 K. The matrix ratios (Ar:HCl) were: (a) 2000:1, (b) 1000:1, (c) 300:1. The broken line in (b) shows the spectrum of the same sample after annealing at 35 K. Spectra (b) and (c) have been offset for clarity. Spectrum (a) shows the monomer absorption with an additional weak band at 2818 cm^{-1} (D), which is assigned to dimeric HCl. The weak band at 2944 cm^{-1} in the monomer spectrum (ν+t) is assigned to the HCl fundamental in combination with a translational mode. In more concentrated matrices (spectra (b) and (c)), bands due to cyclic trimer (cT) and high polymers (P) become prominent. See the text for a discussion. *(Adapted with permission from ref. 2. © 1969, Royal Society of Chemistry.)*

proportion of dimer (Fig. 6.1(a)). The monomer spectrum is not a single line as might be expected, however, but shows discrete rotational structure, which can be analysed as P, Q, and R branches. Close inspection even reveals $^{35}Cl/^{37}Cl$ isotope splittings. The HCl molecule is clearly small enough to undergo some rotation in the matrix cage. There is also a weak band at 2944 cm^{-1} in the monomer spectrum, which is due to the fundamental of HCl coupling with a translational mode of HCl in the Ar lattice, found at 73 cm^{-1}.

In more concentrated matrices, the dimer band becomes more prominent and a strong band due to cyclic trimer appears at 2787 cm^{-1}. At a matrix ratio of 1000:1, the latter band grows appreciably when the matrix is annealed at 35 K to promote diffusion, while additional bands assigned to highly polymeric HCl also grow (Fig. 6.1(b)). At a matrix ratio of 300:1, dimer, cyclic trimer and polymer bands are all prominent, even before annealing, and an extra polymer band is seen at 2701 cm^{-1} (Fig. 6.1(c)).

Repetition of this work may not reproduce the spectra exactly as they appear in Fig. 6.1 at exactly the same matrix ratios, but there should be no major discrepancies. The experiment illustrates some of the matrix effects which one can expect to observe in matrix IR studies. Similar experiments can be carried out with the other hydrogen halides and with different matrix hosts, such as N_2 and CO.[1,2]

1.1.2 Methanol and ethanol in Ar matrices

Methanol is readily available at up to 99.9% purity and containing no more than 0.05% of water; while specially dried methanol, supplied for the preparation of Karl Fischer reagent, typically has a maximum water content of only 0.005%. Ethanol can be obtained in similarly pure form, though usually with slightly higher water content. Deuterium- and ^{13}C-substituted variants of both alcohols can also be purchased at reasonable cost in quantities suitable for matrix experiments. These alcohols are therefore convenient choices for matrix IR experiments.

Studies of the IR spectra of methanol and ethanol in Ar matrices have been made by Barnes and Hallam.[3,4] The $\nu(O–H)$ region for MeOH is particularly interesting, because separate bands can be identified for methanol dimers, trimers, and tetramers, as well as for the monomeric species. The suggested experiment (Protocol 2) is to compare the IR spectra of MeOH in Ar matrices with matrix ratios (Ar:MeOH) of 2000:1 and 50:1.[3]

IR spectra of methanol in Ar matrices at various matrix ratios are shown in Fig. 6.2. At Ar:MeOH = 2000:1, the spectrum consists of a single band due to monomeric MeOH and a set of four weaker bands which are assigned to an open chain dimer. At higher concentrations of MeOH, the relative intensity of the group of dimer bands increases, while other groups, assigned to open chain trimers and tetramers, appear. The broad absorption at 3400–3150 cm^{-1} is due to cyclic tetramers and high polymers of MeOH. The prominent bands expected for monomeric, dimeric, trimeric, and tetrameric MeOH in Ar matrices are summarized in Table 6.1.

Protocol 2.
IR spectra of MeOH in Ar matrices

Equipment

- IR spectrometer
- Calibrated expansion bulb

Materials

- High purity methanol (e.g. specially dried, analytical grade)   **toxic, flammable**
- High purity Ar

Illustrates

- Matrix IR spectroscopy
- Dimer and multimer formation

1. Set up a preparative vacuum line (cf. Figs 4.1 and 4.4) with a 1 litre gas bulb and a cylinder of Ar. Include a calibrated expansion bulb when preparing the gas mixture with matrix ratio 2000:1.

2. Place about 5 ml of methanol in a sample ampoule and degas it as described in Protocol 2 in Chapter 4.

3. Prepare a gas mixture of Ar and methanol with a matrix ratio (Ar:MeOH) of 50:1 or 2000:1, as described in Protocols 3 or 6, Chapter 4, respectively.

4. Transfer the bulb containing the gas mixture to the spray-on line of the cold cell.

5. Following Protocol 7, Chapter 4, deposit the gas mixture as a matrix at 20 K by the slow deposition method.

   NB The original paper gave 20 K as the temperature for deposition, but deposition at the base temperature of the cold cell (e.g. 10–12 K) should work equally well.

6. Record the IR spectrum once a sufficiently thick matrix has been formed.

7. Repeat the experiment with a different matrix ratio. Since the cold window will not need to be cleaned, the cold cell will not need to be opened to the atmosphere after the matrix is boiled off; so it should be feasible to deposit and study both matrices in single day.

At the higher MeOH concentrations, the monomer band develops a splitting, which is probably due to monomer in a second type of matrix site. The matrix cage containing a trapped MeOH molecule will be distorted by MeOH molecules which are in the vicinity, but which are not close enough to form H-bonds with the MeOH molecule in the cage.

For a full analysis of the IR spectra of MeOH in Ar matrices, including C–H stretching and all other regions, the reader should consult the original paper.[3]

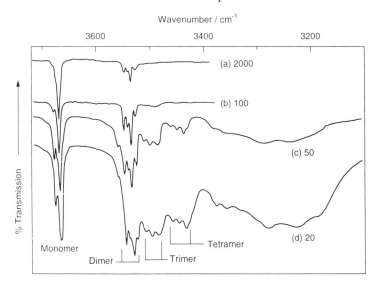

Fig. 6.2 IR spectra in the ν(O–H) region of MeOH in Ar matrices at 20 K. Matrix ratios (Ar:MeOH) were: (a) 2000:1, (b) 100:1, (c) 50:1, (d) 20:1. Spectra (b)–(d) have been offset for clarity. See the text for a discussion and Table 6.1 for a summary of the wavenumber data. *(Adapted with permission from ref. 3. © 1970, Royal Society of Chemistry.)*

Table 6.1. Prominent ν(OH) bands of MeOH in Ar matrices.[a]

Wavenumber / cm^{-1}	Assignment
3679	Monomer—second site?
3667	Monomer
3541	
3533	Open chain dimer
3528	
3519	
3505	
3495	Open chain trimer
3482	
3458	
3446	Open chain tetramer
3435	

[a] Data and assignments from ref. 3.

The experiment can be extended to include studies of ethanol and deuteriated variants of both alcohols.[3,4]

1.2 Complexes of stable molecules in matrices

In most cases, the study of complexes of stable molecules in matrices requires the simultaneous deposition of two matrix-gas mixtures. The simple spray-on line shown in Fig. 4.5 is inadequate for this purpose; essentially two such lines will be needed, preferably feeding into separate inlet ports on the vacuum shroud of the cold cell. Since it is also desirable to spray on the two gas mixtures at identical rates, high-quality needle valves that have been calibrated for rates of gas flow (cf. Chapter 4, Section 3.3.1) should be used if at all possible.

1.2.1 The 1:1 water–ammonia complex in N_2 matrices

The so-called linear 1:1 complex formed between water and ammonia (Fig. 6.3) has been studied by IR spectroscopy in N_2 and Ar matrices by Nelander and Nord[5] and by Yeo and Ford;[6] and in Ne and Kr matrices by Engdahl and Nelander.[7] Anhydrous ammonia of better than 99.98% purity can be bought in small cylinders, and is ideal for this experiment, although quite expensive in this form. The suggested experiment (Protocol 3) is to prepare ammonia–N_2 and water–N_2 mixtures, each with a matrix ratio (N_2:guest) of 100:1, co-deposit these as a matrix at 17 K, and examine the IR spectra before and after annealing.[6] Note that if the two matrix-gas mixtures are deposited simultaneously at the same rate, the final matrix ratio should be $N_2:H_2O:NH_3 = 200:1:1$.

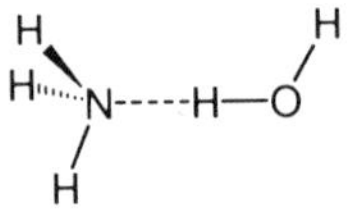

Fig. 6.3 The linear 1:1 complex of water and ammonia.

Protocol 3.
IR spectra of the water–ammonia complex in N_2 matrices

Caution! NH_3 is toxic and should not be allowed to escape from the cylinder into the open laboratory. Fit the correct type of stainless steel valve to the cylinder, store the cylinder in a fume cupboard or other well ventilated place when it is not in use, and vent any excess NH_3 into a fume cupboard or to the outside of the building.

Equipment
- Cold cell with a temperature controller, heater and two spray-on lines
- IR spectrometer

Materials
- Distilled water
- High purity ammonia (cylinder)   **toxic, flammable**
- High purity N_2

Illustrates
- Matrix IR spectroscopy
- Co-deposition of two matrix gases
- Annealing of matrices
- Complex formation

1. Set up a preparative vacuum line (cf. Fig. 4.1) with a 1 litre gas bulb and cylinders of Ar and NH_3. (See Chapter 4, Section 2.4.4 for advice on connecting the two cylinders.)

2. Prepare a gas mixture of N_2 and NH_3 with $N_2:NH_3 = 100:1$, following Protocol 3, Chapter 4.

3. Transfer the bulb containing the gas mixture to one of the spray-on lines of the cold cell.

4. Place about 5 ml of distilled water in a sample ampoule and degas it as described in Protocol 2, Chapter 4.

5. Prepare a gas mixture of N_2 and H_2O with a matrix ratio ($N_2:H_2O$) of 100:1, as described in Protocol 3, Chapter 4.

6. Transfer the bulb containing the gas mixture to the second spray-on line of the cold cell.

7. Following Protocol 7, Chapter 4, deposit both gas mixtures simultaneously as a matrix at 17 K by the slow deposition technique. Try to adjust the needle valves of the two spray-on lines to give identical spray-on rates for both gas mixtures; this will need careful monitoring of the pressure drop in each line.

   NB The original paper gave 17 K as the temperature for deposition, but deposition at the base temperature of the cold cell (e.g. 10–12 K) should work equally well.

8. Record the IR spectrum once a sufficiently thick matrix has been formed.

9. Anneal the matrix briefly at 30 K as described in Instruction 5 of Protocol 1.

10. Record the IR spectrum again.

11. The experiment may be continued with further periods of annealing, recording an IR spectrum after each one. Both the duration and temperature of annealing may be increased, although matrix boil-off will become excessive above about 35 K.

Under the conditions of the experiment, there does not seem to be much of the complex formed initially upon matrix deposition. When the matrix is annealed to allow diffusion, however, the 1:1 water–ammonia complex is formed in appreciable amounts, as shown by the growth of its IR bands. Dimers of water and ammonia and higher multimers and mixed multimers, are also formed. Figure 6.4 shows the $v(O–H)$ and $v(N–H)$ regions of the IR spectrum of an N_2 matrix containing H_2O and NH_3, before and after annealing. The spectrum initially consists principally of the IR bands of monomeric H_2O and NH_3, but, after annealing, bands due to the complex and dimeric H_2O and NH_3 have all grown in. Figure 6.5 shows a similar pattern for the symmetric bending mode (v_2) of NH_3 and its equivalents in the complex and ammonia dimers.

Full discussions of the water–ammonia complex in matrices, including

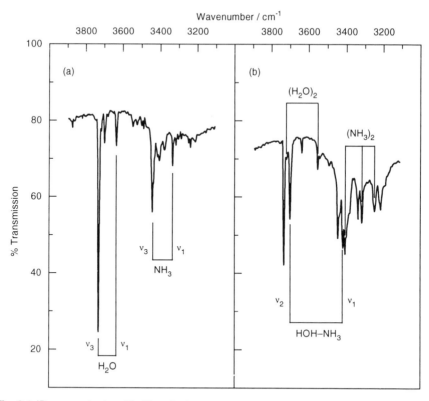

Fig. 6.4 IR spectra in the ν(O–H) and ν(N–H) regions of water and ammonia co-deposited in N_2 at 17 K (N_2:H_2O:NH_3 = 200:1:1): (a) before and (b) after annealing at 30 K. Bands due to the complex HOH–NH_3 and dimers, $(H_2O)_2$ and $(NH_3)_2$, can be seen to have grown on annealing. *(Adapted with permission from ref. 6. © 1991, Elsevier.)*

explanations of how the various species were identified, can be found in the original literature.[5-7]

1.2.2 Amine–hydrogen chloride complexes

Ammonia and alkyl amines form strongly H-bonded complexes with the hydrogen halides, and these have been studied by matrix IR spectroscopy since the early 1970s. Ault and Pimentel were the first to record the IR spectrum of the matrix isolated ammonia–HCl complex,[8] and this work was later extended to include the trimethylamine–HCl complex and also complexes with HBr.[9] From the observed HCl frequencies it was conclude that the proton is shared more or less equally between the basic nitrogen and the halogen in the ammonia-HCl complex, but in the $(CH_3)_3N$–HCl complex the proton is more closely bonded to the base, tending towards the ionic structure, $(CH_3)_3NH^+Cl^-$. Later, Barnes *et al.* investigated these and a range of related complexes and discovered striking differences between the IR spectra in Ar

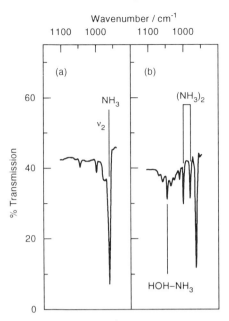

Fig. 6.5 IR spectra in the region 1150–900 cm^{-1} of water and ammonia co-deposited in N_2 at 17 K (N_2:H_2O:NH_3 = 200:1:1): (a) before and (b) after annealing at 30 K. Bands due to the complex HOH–NH_3 and ammonia dimer grew on annealing. *(Adapted with permission from ref. 6. © 1991, Elsevier.)*

and N_2 matrices.[10] It appears that proton transfer from hydrogen chloride to the amine is increased in the more polarizable nitrogen matrix. The suggested experiment (Protocol 4) is to prepare either the methylamine–HCl or the trimethylamine–HCl complex in both an Ar and an N_2 matrix and compare the IR spectra. Anhydrous methylamine and trimethylamine are available in small cylinders with purities of at least 98%.

Protocol 4.
IR spectra of the methylamine-HCl or trimethylamine-HCl complex in Ar and N_2 matrices

Caution! The amines are flammable, harmful by inhalation, and irritating to the eyes, respiratory system, and skin. They should not be allowed to escape from the cylinder into the open laboratory. Store cylinders in a fume cupboard or other well ventilated place when they are not in use and vent any excess amine into a fume cupboard or to the outside of the building. Stainless steel valves are recommended. Note also the caution regarding HCl in Protocol 1.

Equipment
- Cold cell with two spray-on lines
- IR spectrometer

185

Protocol 4. *Continued*

Materials

- Methylamine or trimethylamine at least 98% pure (cylinder)   **irritant, flammable**
- HCl of at least 99% purity (cylinder)   **irritant, corrosive**
- High purity Ar
- High purity N_2

Illustrates

- Matrix IR spectroscopy
- Co-deposition of two matrix gases
- Complex formation
- Matrix host effects

1. Set up a preparative vacuum line (cf. Fig. 4.1) with a 1 litre gas bulb and cylinders of Ar and the amine. (See Chapter 4, Section 2.4.4 for advice on connecting the two cylinders.)

2. Prepare a gas mixture of Ar and the amine with Ar:amine = 100:1, following Protocol 3, Chapter 4.

3. Transfer the bulb containing the gas mixture to one of the spray-on lines of the cold cell.

4. Prepare a mixture of Ar and HCl with Ar:HCl = 100:1, following the first part of Protocol 1 (this chapter) and Protocol 3, Chapter 4.

5. Transfer the bulb containing the Ar–HCl mixture to the second spray-on line of the cold cell.

6. Following Protocol 7, Chapter 4, deposit both gas mixtures simultaneously as a matrix at 20 K by the slow deposition technique. Try to adjust the needle valves of the two spray-on lines to give identical spray-on rates for both gas mixtures; this will need careful monitoring of the pressure drop in each line. Equal rates of deposition will result in a matrix with an overall matrix ratio of Ar:amine:HCl = 200:1:1.

   NB The original paper gave 20 K as the approximate temperature for deposition, but deposition at the base temperature of the cold cell (e.g. 10–12 K) may work equally well.

7. Record the IR spectrum once a sufficiently thick matrix has been formed.

8. Repeat the experiment with N_2 as the matrix host.

Figure 6.6 shows IR spectra obtained with Ar and N_2 matrices containing methylamine and hydrogen chloride. Figure 6.7 gives a similar comparison for trimethylamine and hydrogen chloride. To interpret the results, matrix IR spectra of methylamine and trimethylamine are needed for comparison. These have already been published,[11-13] but it might be as well to record them afresh with the actual cold cell and spectrometer to be used for this experiment. In Figs 6.6 and 6.7, the absorptions belonging to the uncomplexed amine are shown as broken lines.

It is immediately apparent from the differences between the Ar and N_2 matrix spectra that the host material exerts a profound effect. Nitrogen, being

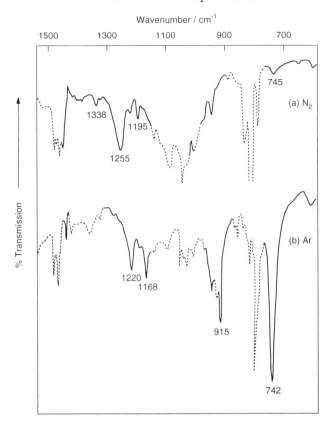

Fig. 6.6 IR spectra of matrices containing methylamine and hydrogen chloride. (a) N_2 matrix; (b) Ar matrix (offset for clarity). The matrices were deposited at about 20 K, with matrix ratios (host:amine:HCl) of 200:1:1. The portions of the spectra shown as broken lines show the absorptions of uncomplexed methylamine. *(Adapted with permission from ref. 10. © 1984, Royal Society of Chemistry.)*

more polarizable than argon, appears to favour a greater degree of proton transfer from HCl to the basic nitrogen of the amine. The IR bands of the complex which are particularly sensitive to this difference are the asymmetric stretch and bending modes of the N···H···Cl grouping. The wavenumber data for these modes are summarized in Table 6.2.

For methylamine, the strong band at 742 cm^{-1} in the Ar matrix spectrum (Fig. 6.6(b)) is assigned to the N···H···Cl asymmetric stretching mode of the 1:1 methylamine–HCl complex. At first sight, it appears to have no counterpart in the N_2 matrix spectrum (Fig. 6.6(a)). Closer scrutiny, however, reveals a very broad absorption, upon which other sharper bands are superimposed, with a maximum near 1050 cm^{-1}; this is assigned as the asymmetric N···H···Cl mode, shifted by about 300 cm^{-1} from its position in the Ar matrix spectrum.

Trimethylamine has a greater proton affinity than methylamine, and the

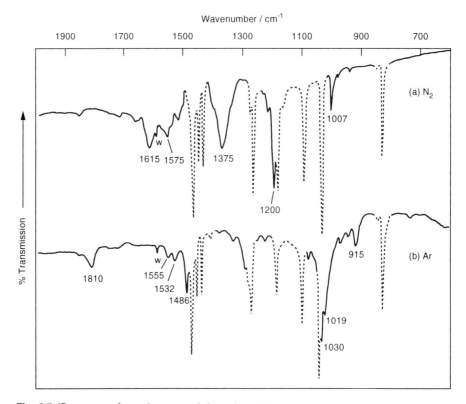

Fig. 6.7 IR spectra of matrices containing trimethylamine and hydrogen chloride. (a) N_2 matrix; (b) Ar matrix (offset for clarity). The matrices were deposited at about 20 K, with matrix ratios (host:amine:HCl) of 200:1:1. The portions of the spectra shown as broken lines show the absorptions of uncomplexed trimethylamine. The small band marked w is due to water impurity. *(Adapted with permission from ref. 10. © 1984, Royal Society of Chemistry.)*

N⋯H⋯Cl asymmetric stretching mode of the 1:1 trimethylamine–HCl complex appears at higher frequencies than the equivalent mode of CH_3NH_2–HCl; being 1486 and 1615 cm^{-1} in Ar and N_2, respectively. The prominent broad absorption at 1375 cm^{-1} in the N_2 matrix spectrum (Fig. 6.7(a)) is assigned to the N⋯H⋯Cl bend.

In view of the large differences between Ar and N_2 matrix spectra, reliable assignment of the observed IR bands in these experiments was not completely straightforward, but was greatly assisted by deuterium substitution. Comparable results were also obtained for complexes of HCl with ethylamine and dimethylamine. A detailed discussion can be found in the original paper.[10]

1.2.3 The dimethylformamide–hydrogen chloride complex

If the reader wishes to carry out a matrix experiment involving complex formation, but does not wish to incur the expense of buying cylinders of

Table 6.2. Antisymmetric stretching and bending modes (cm^{-1}) of the N···H···Cl grouping in amine–HCl complexes in Ar and N$_2$ matrices.[a]

Amine	Ar matrix		N$_2$ matrix	
	v_{as}	δ	v_{as}	δ
Methylamine	742	1168	~1050	1255
				1195
Trimethylamine	1486	1030	1615	1375
		1019		

[a] Data and assignments from ref. 10.

ammonia or alkyl amines, a study of the DMF–HCl complex provides an inexpensive and easy option. DMF can be purchased either as HPLC grade, with 99.9% purity and water content less than 0.03%, or as the anhydrous material with less than 0.005% water. In the original experiments, described by Mielke *et al.*,[14] gas-phase mixtures of *(i)* Ar or N$_2$ and DMF and *(ii)* Ar or N$_2$ and HCl were made up (cf. Protocol 3, in Chapter 4 and Protocol 1, this chapter), and these were co-deposited at about 20 K, as described in the previous section, to yield matrices with a variety of DMF and HCl relative concentrations. The published paper shows the IR spectrum of a nitrogen matrix with matrix ratio N$_2$:HCl:DMF = 500:4:1, and this would be a good choice for making comparative tests.

2. Reactive species in matrices

2.1 Small reactive molecules and fragments

2.1.1 Nitrogen atoms

If a small microwave generator is available, one of the simplest matrix experiments which can be carried out—and yet one which is still a good test of the ability of a matrix to stabilize very reactive species—is the trapping of nitrogen atoms in a nitrogen matrix. In this experiment (Protocol 5), nitrogen is both the precursor for the reactive species and the host material. The successful trapping of N atoms can be confirmed visually, by warming the matrix and observing the green luminescence which arises from the excited state of N$_2$ that is formed initially when two N atoms recombine (Scheme 6.1).

N$_2$ $\xrightarrow[\text{2. Trapping at 10–20 K}]{\text{1. Microwave discharge}}$ N + N $\xrightarrow{\Delta}$ [N$_2$]* $\longrightarrow$ N$_2$ + $h\nu$

excited   ground
state   state

Scheme 6.1

189

Thus, there is no need for a spectrometer in this experiment, but if an ESR spectrometer and suitable cold cell (cf. Fig. 2.8) are available, the nitrogen atoms can also be observed directly by ESR.

Protocol 5.
The generation and trapping of nitrogen atoms in a nitrogen matrix

Equipment
- Microwave generator
- ESR spectrometer (optional)
- Cold cell with sample rod and shroud suitable for ESR spectroscopy (optional)

Materials
- High purity nitrogen

Illustrates
- Generation of reactive species by microwave discharge
- Matrix ESR spectroscopy (optional)

1. The experiment is best carried out in a darkened laboratory, so that the luminescence can be seen clearly. It is not necessary to have a full black-out, however.

2. Set up the cold cell with a Pyrex inlet tube of about 15 mm o.d. and the cavity of the microwave generator surrounding the tube about 10 cm away from the shroud. Figure 6.8 shows the arrangement with an ESR shroud. If an ESR shroud is used, make sure that the flexible connection between the spray-on line and the Pyrex inlet tube allows enough vertical movement for the lower part of the shroud to be raised fully for recording ESR spectra (cf. Fig. 2.8).

3. Fill a 1 litre gas bulb to about 800 mbar with high purity nitrogen, as described in Protocol 1, Chapter 4.

4. Transfer the bulb to the spray-on line of the cold cell.

5. Following Protocol 7, Chapter 4, begin depositing the nitrogen as a matrix at 10–20 K by the slow deposition method. The rate of deposition is not critical; 5 mmol h^{-1} should be about right.

6. Once a steady rate of deposition has been established, switch on the microwave generator, and adjust the power output and tune the cavity to give a bright glow in the pyrex inlet tube. The excited gas should glow with a white or pinkish-white luminescence.

7. Observe the cold window (or cold rod if an ESR set-up is employed). There should be a discernible green luminescence coming from the accumulating matrix, indicating the formation of N atoms.

8. Continue depositing nitrogen through the microwave discharge for about 20–30 minutes.

9. Switch off the microwave generator. The green glow on the cold window (or cold rod) will decrease in intensity over a number of seconds, then disappear.

10. At this stage, if equipped for this, record an ESR spectrum of the trapped N atoms.

11. Warm the matrix, either by using the heater and temperature controller (cf. instruction 5 of Protocol 1) or by switching off the helium refrigerator.

12. Observe the cold window (or rod) visually. When diffusion of the N atoms becomes significant, at about 15–25 K, the green luminescence will re-appear. Depending on the rate of heating, the glow may continue at moderate intensity for many seconds, or it may appear as a brief flash.

13. With a little planning, the chemiluminescence can be photographed, and makes an attractive picture for lectures and presentations.

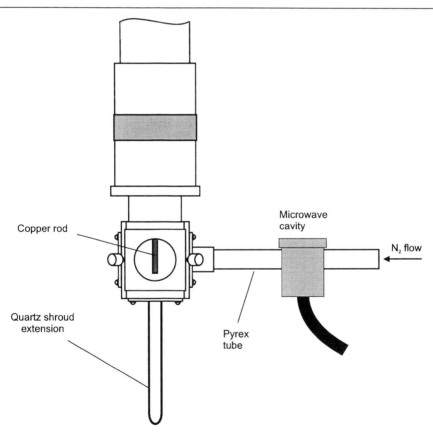

Fig. 6.8 Generation of nitrogen atoms and observation by ESR spectroscopy. The ESR shroud is fully extended for spray-on (cf. Fig. 2.8). With N_2 flowing and the microwave generator switched on, a glow discharge will appear, extending a few centimetres each side of the cavity. The matrix accumulating on the cold copper rod will have a green luminescence due to N-atom recombination.

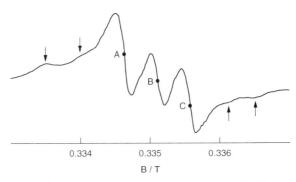

Fig. 6.9 ESR spectrum of nitrogen atoms trapped in N_2 at 4.2 K. Nitrogen atoms were generated by conducting N_2 gas through a microwave cavity, as shown in Fig. 6.8. The spectrum was recorded at 9397 MHz. The three components of the main 1:1:1 triplet are shown as A, B, and C. The arrows indicate weak satellite signals each side of the main signal, due to interaction between N atoms and neighbouring N_2 molecules. *(Adapted with permission from ref. 17. © 1957, American Institute of Physics.)*

The green emission from nitrogen contributes to the aurora borealis and has thus been observed since before recorded history, as an object of mystery and romance. In the laboratory, Vegard was the first to study the emission from nitrogen at low temperatures, which he observed during and following the bombardment of solid nitrogen with electrons and positive ions.[15] Matrix-isolation studies of the emission spectra arising from N-atom recombination[16] and the matrix ESR spectra of N atoms themselves[17,18] were both reported fairly soon after the original development of the matrix-isolation technique. The ESR signal of matrix isolated N atoms is a 1:1:1 triplet due to the electron coupling with a single ^{14}N nucleus with nuclear spin $I = 1$ (Fig. 6.9). Weak satellite triplets can also be seen, partially overlapping the main signal at each side. These arise from interactions between N atoms and neighbouring N_2 molecules.

2.1.2 Nitroxyl (or nitrosyl hydride), HNO

Nitroxyl (HNO) was one of the first unstable species to be positively identified and characterized by matrix IR spectroscopy. Brown and Pimentel[19] investigated the photolysis of nitromethane and methyl nitrite in Ar matrices and established that, under these conditions, nitromethane simply isomerizes to methyl nitrite, and all other products come from secondary photolysis of the nitrite. The primary photoproducts of methyl nitrite are formaldehyde and nitroxyl (Scheme 6.2). Brown and Pimentel identified two IR absorptions, at 1570 and 1110 cm^{-1}, as belonging to this species, but the latter was eventually

$$CH_3{-}NO_2 \xrightarrow{h\nu} CH_3{-}ONO \xrightarrow{h\nu} HCHO + HNO$$

Scheme 6.2

shown to be wrongly assigned.[20] None the less, this was the first time that the IR spectrum of HNO had been observed. Later, the matrix photolyses of methyl nitrite[20] and nitromethane[21] were studied in greater detail and it became clear that HNO and formaldehyde were generated as an H-bonded pair. For this complex, the N⋯H⋯O stretching mode gave rise to an IR absorption at 2802 cm^{-1}.

Nitromethane (a liquid) can be purchased inexpensively with a purity of at least 99% and is the most convenient starting material for this experiment. Methyl nitrite, on the other hand, is a gas, which is usually prepared as needed by the action of sulphuric acid on methanol and sodium nitrite.[22,23] Methyl nitrite prepared in this way can be purified by trap-to-trap distillation on a vacuum line, but the suggested experiment (Protocol 6) is to take the route of convenience, and photolyse nitromethane in an Ar matrix, as described by Jacox,[21] and record the IR spectrum of the resulting matrix.

Protocol 6.
The matrix photolysis of nitromethane and the IR spectrum of HNO

Caution! Note that striking an arc lamp too close to sensitive electronic equipment can result in malfunction or damage (see Chapter 3, Section 2.1.1, subsection *i*).

Equipment

- IR spectrometer
- Calibrated expansion bulb

- Hg arc lamp (medium or high pressure) with a water filter

Materials

- Nitromethane (preferably of 99+% purity)   **harmful, flammable**
- High purity Ar

Illustrates

- Matrix IR spectroscopy
- Matrix photolysis

1. Set up a preparative vacuum line (cf. Fig. 4.4) with a 1 litre gas bulb, a cylinder of Ar, and a calibrated expansion bulb.

2. Place about 5 ml of nitromethane in a sample ampoule and degas it as described in Protocol 2, Chapter 4.

3. Prepare a gas mixture of Ar and nitromethane with a matrix ratio (Ar:CH_3NO_2) of between 500:1 and 1000:1, as described in Protocol 6, Chapter 4.

4. Transfer the bulb containing the gas mixture to the spray-on line of the cold cell.

5. Following Protocol 7, Chapter 4, deposit the gas mixture as a matrix at 10–20 K by the slow deposition method.

Protocol 6. *Continued*

6. Record the IR spectrum once a sufficiently thick matrix has been formed.

7. Photolyse the matrix with a mercury arc fitted with a water filter (cf. Fig. 3.3), monitoring the progress of the reaction by recording IR spectra at intervals.

8. Look for the IR bands of HNO at 2802 (ν(N–H) H-bonded to HCHO) and 1570 cm^{-1} (ν(NO)), and also the ν(CO) band of formaldehyde at about 1740 cm^{-1}. The total photolysis time required to generate a reasonable amount of HNO will vary with the lamp power and type of light filtration used, but can be expected to be about one to six hours.

9. The experiment can be varied by carrying out photolysis simultaneously with deposition, though in these circumstances an IR spectrum of the unphotolysed nitromethane is not obtained.

With nitromethane as the starting material, the primary photochemical process will be the generation of methyl nitrite. A matrix IR spectrum of methyl nitrite is therefore needed for comparison; two thorough IR studies of methyl nitrite in Ar matrices have been published.[24,25] In the IR spectrum of the matrix-isolated nitrite, bands due to both *cis* and *trans* rotamers can be distinguished. Jacox's paper[21] gives a full list of the IR bands observed after 35, 163, and 396 minutes photolysis of nitromethane in Ar matrices; this paper should certainly be consulted when the experiment is attempted.

2.1.3 The formyl radical, HCO

The formyl radical was first detected by its visible absorptions in the gas phase.[26] Subsequently, it was one of the first radical species to be characterized by matrix IR spectroscopy, when Ewing *et al.* photolysed HI and HBr in CO matrices.[27] Two IR bands belonging to HCO were identified: ν_2 (the C=O stretch) at 1860 cm^{-1} and ν_3 (the bend) at 1091 cm^{-1}. The ν_1 band (C-H stretch) was not assigned in this first study, but was later identified by Milligan and Jacox[28] at the unusually low frequency of 2488 cm^{-1}. The assignment of the fundamental vibrations of HCO has been discussed in detail by Ogilvie.[29]

At about the same time as the matrix IR studies, the ESR spectrum of the formyl radical in CO matrices was also observed by Adrian *et al.*,[30] thus giving structural information which complemented that obtained from IR spectroscopy. More recently, it was shown that photolysis of matrix-isolated HCO with plane-polarized yellow light results in photo-reorientation of the radical and an anisotropic ESR signal.[31]

Anhydrous gaseous HBr can be purchased in small cylinders (with >99% purity). It may also be possible to find a supplier of gaseous HI, but this hydrogen halide is much more widely available as a concentrated aqueous solution. Anhydrous gaseous HI can be obtained from the aqueous solution by adding P_2O_5, then passing the liberated gas through a tube packed with P_2O_5, and condensing it in a cooled sample ampoule (cf. Fig. 4.1). The liquid HI (m.p. $-51\,°C$; b.p. $-35\,°C$) should finally be purified by several trap-to-trap distillations on the vacuum line, and kept cold and away from light until needed. The suggested experiment (Protocol 7) is to prepare a matrix-gas mixture of CO and either HBr or HI (CO:hydrogen halide = 500:1), deposit this as a matrix, and record IR spectra before and after photolysis and also after a final annealing. The experiment may be varied, however, to include visible or ESR spectroscopy, and photo-orientation of the formyl radical with plane-polarized light.

i. The IR spectrum of HCO
Protocol 7 follows fairly closely the original experiments of Ewing *et al.*[27]

Protocol 7.
The matrix photolysis of HBr or HI in a CO matrix: the IR spectrum of HCO

Caution! HBr and HI are irritants and corrosive, and should not be allowed to escape from the cylinder into the open laboratory. Fit the correct type of stainless steel valve to the cylinder, store the cylinder in a fume cupboard or other well ventilated place when it is not in use, and vent any excess hydrogen halide into a fume cupboard or to the outside of the building. When making up the matrix-gas mixture, avoid exposing expensive and delicate gauge heads to the hydrogen halide; use an inexpensive Bourdon gauge or a manometer (cf. Fig. 3.2) instead.

Note also that striking an arc lamp too close to sensitive electronic equipment can result in malfunction or damage (see Chapter 3, Section 2.1.1, subsection *i*).

Equipment
- Cold cell with a heater and temperature controller
- IR spectrometer
- Calibrated expansion bulb
- Medium- or high-pressure Hg arc with a water filter

Materials
- HBr or HI (in a cylinder or gas bulb)   **irritant, corrosive**
- High purity CO   **toxic, flammable**

Illustrates
- Matrix IR spectroscopy
- Matrix photolysis
- Reactive matrix host
- Annealing matrices

1. Set up a preparative vacuum line (cf. Fig. 4.4) with a 1 litre gas bulb, calibrated expansion bulb, and cylinders of CO and HBr or HI. (See Chapter 4, Section 2.4.4 for advice on connecting the two cylinders.) Alternatively, if the hydrogen halide is in a gas bulb, mount this on a vacant port on the vacuum line.

Protocol 7. *Continued*

2. Wrap the gas bulb with aluminium foil, to minimize exposure of the hydrogen halide to light.

3. Prepare a gas mixture of CO and the hydrogen halide (HX) with a matrix ratio (Ar:HX) of about 500:1, as described in Protocol 6, Chapter 4.

4. Transfer the bulb containing the gas mixture to the spray-on line of the cold cell.

5. Following Protocol 7, Chapter 4, deposit the gas mixture as a matrix at 20 K by the slow deposition technique.

   NB The original paper gave 20 K as the temperature for deposition, but deposition at the base temperature of the cold cell (e.g. 10–12 K) should work equally well.

6. Record the IR spectrum once a sufficiently thick matrix has been formed.

7. Photolyse the matrix with a mercury arc fitted with a water filter (cf. Fig. 3.3), monitoring the progress of the reaction by recording IR spectra at intervals.

8. Look for the IR bands of HCO at 2488, 1860, and 1091 cm^{-1}. The total photolysis time required to generate a reasonable amount of HCO will vary with the lamp power, but can be expected to be only a few minutes.

9. Anneal the matrix briefly at 45 K as described in Instruction 5 of Protocol 6.1.

10. Record the IR spectrum again.

Figure 6.10 shows IR spectra recorded in the original work with HI in CO matrices.[27] Similar spectra were obtained with HBr. On photolysis of the hydrogen halide, H atoms are generated, and these react with CO to give the formyl radical, with IR bands at 2488, 1860 and 1091 cm^{-1} (Fig. 6.10(b)). The other prominent photoproduct band in Fig. 6.10b at 1730 cm^{-1} is due to

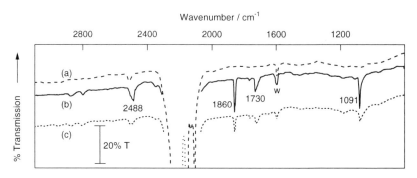

Fig. 6.10 Photolysis of HI in a CO matrix. IR spectra of HI in solid CO at 20 K (CO:HI = 440:1): (a) before photolysis; (b) after 8 minutes photolysis; (c) after annealing at 45 K. Spectra (b) and (c) are offset for clarity. The band marked w is due to water impurity. *(Adapted with permission from ref. 27. © 1960, American Institute of Physics.)*

formaldehyde, while the weak band at 1760 cm^{-1} is assigned to the dimer of the formyl radical, glyoxal. On annealing the matrix at 45 K, the bands due to HCO diminish in intensity, as expected when a reactive species is allowed to diffuse (Fig. 6.10(c)), but it is not clear from the spectrum what products result from this thermal process. In the original work, identification of the formyl radical was supported by experiments with DBr and DI, from which DCO was generated, and by force-field calculations. The reader should consult the earlier papers for a complete description.[27-29]

ii. Visible absorption spectroscopy of HCO

With a UV–visible spectrometer replacing the IR spectrometer, Protocol 7 can be followed to observe the visible absorptions of HCO. Ewing *et al.* reported seven absorption maxima at 510, 533, 555, 579, 605, 635, and 670 nm for this radical in a CO matrix.[27] If the facility is available to move the cold cell from a UV–visible spectrometer to an IR spectrometer without losing the matrix, it should be possible to combine the observation of both visible and IR absorptions in a single experiment.

For a combined IR and UV–visible experiment, use a cold cell with a CsBr or CsI cold window and KBr external windows. Begin by taking background spectra in both the UV–visible and IR regions, then carry out the UV–visible study with a thin matrix and short photolysis times. Continue the experiment by spraying on more of the matrix-gas mixture and subjecting the matrix to further periods of photolysis. UV–visible spectra and IR spectra can be recorded at each stage, but the matrix is likely to become opaque in the UV, through both scattering and absorption, as its thickness builds up.

For a simple UV–visible experiment, it would be better to use a CaF$_2$ or sapphire cold window and quartz outer windows.

iii. ESR spectroscopy of HCO

For those wishing to carry out a matrix ESR experiment with HCO, it is necessary to have a cold cell which is suitable for ESR (cf. Fig. 2.8), as well as an ESR spectrometer. Protocol 7 can then be followed to obtain an ESR spectrum of the HCO radical. Figure 6.11 shows that this consists of two non-symmetrical lines spaced about the free electron g value and separated by about 13.6 mT.[30] The experiment can be elaborated by investigating the polarized photolysis of the HCO radical, which has been shown to produce an oriented sample with an anisotropic ESR spectrum.[31] Photo-reorientation of HCO molecules can be induced with plane-polarized yellow light ($\lambda > 500$ nm). If the plane of polarization is chosen so that the photolysis beam has its $\boldsymbol{E}$-vector parallel to the applied magnetic field of the ESR spectrometer, the resulting photo-reorientation shows up as changes in the ESR spectrum when the rod on which the matrix is deposited is rotated 90° about its axis. The original workers called this phenomenon *magnetophotoselective photolysis*, but it is the same as photo-reorientation discussed in Chapter 5, Section 1.5,

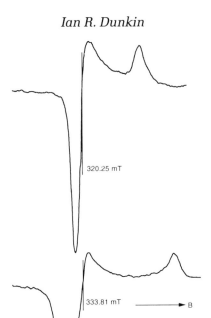

320.25 mT

333.81 mT ——————→ B

1 mT

Fig. 6.11 ESR signals of the formyl radical, HCO, in a CO matrix at 11 K. HCO was generated by photolysis of HI in the CO matrix. *(Adapted with permission from ref. 30. © 1963, American Institute of Physics.)*

except that detection of orientation is by ESR rather than IR or UV–visible spectroscopy. The HCO radicals are reoriented by photodissociation into H atoms and CO molecules, followed by recombination of the H atoms at random with either the same or different CO molecules. This experiment with polarized light and ESR spectroscopy requires only a visible polarizer, such as a plastic film or photographic polarizer, rather than a more expensive quartz or prism type, and could therefore be a convenient exercise in polarized photolysis. The anisotropy achieved in the original work, however, although quite real, was also quite small, and became clear only in magnified difference spectra.

2.1.4 The methyl radical from pyrolysis of methyl iodide

The first attempts to generate the methyl radical in matrices and observe its IR spectrum included reaction of methyl iodide or bromide with lithium atoms[32] and vacuum-UV photolysis of methane.[33] Although the radical was apparently formed in these processes, there were uncertainties and discrepancies in the results. The first success in observing all three IR-active vibrations of the matrix-isolated methyl radical came in experiments by Snelson, in which the radical was generated by gas-phase pyrolysis of methyl iodide and dimethylmercury.[34]

This was one of the first successful demonstrations of the use of gas-phase pyrolysis in conjunction with matrix isolation. The original work made use of a platinum pyrolysis tube about 4.5 cm long with an i.d. of about 2.3 mm, which could be heated resistively to 1700 K. The pyrolysis zone was packed with a small amount of platinum gauze, to achieve better heat transfer. The methyl radical precursor was allowed into the pyrolysis tube at a rate of about 0.01–0.05 mmol h^{-1}, while the matrix host gas was separately deposited at about 50 mmol h^{-1}.

It may be worth trying this experiment with a pyrolysis tube made of alumina or other refractory material (see Chapter 3, Section 2.2), provided it can be heated to at least 1300 K. Neon was used as the host gas in the original experiments, but argon should be substituted if the base temperature of the cold cell is 10 K or above. Set up a cold cell equipped with two spray-on lines. Install a gas bulb containing the matrix host gas on one of the lines, which should feed directly into the shroud via a needle valve. Connect the other line via the pyrolysis tube to the shroud, and install on it a second gas bulb containing methyl iodide vapour. For matrix deposition, heat the pyrolysis tube to 1300–1700 K, begin by depositing pure host gas at about 50 mmol h^{-1}, then bleed methyl iodide into the pyrolysis tube at about 0.05 mmol h^{-1}. Some preliminary experiments to establish the needle valve setting for this latter rate of spray-on may help, and it would be easier to monitor such a slow rate of gas flow if an oil or dibutyl phthalate manometer (cf. Fig. 3.2) were used to record the pressure drop, rather than a less sensitive type of gauge. Monitor the IR spectrum of the developing matrix.

It is likely to take several hours of pyrolysis and deposition to attain sufficient intensity in the matrix IR bands. Table 6.3 lists the IR absorptions that were observed for a neon matrix, which include bands of unreacted CH_3I,

---

Table 6.3. IR bands observed after pyrolysis of CH_3I at 1600 K and trapping the products in a neon matrix.[a]

Wavenumber/cm^{-1}	Species	Wavenumber/cm^{-1}	Species
3162	CH_3	1467	C_2H_6
3055	CH_3I	1435	CH_3I
3015	CH_4	1409	CH_3I
2979	C_2H_6	1398	CH_3
2965	CH_3I	1305	CH_4
2948	C_2H_6	948	C_2H_4
2921	C_2H_6	883	CH_3I
2888	C_2H_6	819	C_2H_6
2855	CH_3I	614	CH_3
2830	CH_3I	532	CH_3I

[a] Data and assignments from ref. 34.

ethane, ethylene and methane, as well as the three IR bands of the methyl radical. The frequencies in an argon matrix will not differ greatly from those in neon. The methyl radical band at 614 cm^{-1} should be relatively intense, but those at 3162 and 1398 cm^{-1} will be weak. In the original work, assignment of the methyl bands was supported by additional pyrolysis experiments with CD_3I, and $Hg(CH_3)_2$.

2.1.5 The trichloromethyl radical from reaction of lithium with carbon tetrachloride

The co-deposition of lithium atoms effusing from a Knudsen cell (cf. Figs 3.6 and 3.7) with a matrix-gas mixture containing an alkyl halide generally results in a reaction to yield the corresponding lithium halide, together with the alkyl radical and products derived from it. The reaction takes place either in the gas phase just prior to deposition, or on the surface of the matrix before complete cooling has occurred. One of the classic demonstrations of this method of generating reactive species in matrices, first reported by Andrews,[35] is the reaction of carbon tetrachloride with lithium to give the trichloromethyl radical. IR bands of the carbon–chlorine stretching vibrations of CCl_3 were observed at 898 and 674 cm^{-1}. Since the starting materials are readily available, a repeat of the simplest of these experiments makes a good test of a cold cell equipped with a Knudsen furnace.

Carbon tetrachloride is available as HPLC grade with better than 99.9% purity, or as the anhydrous material (99%) with less than 0.005% water. Natural lithium consists of 7.42% of ^6Li and 92.58% of ^7Li. In the original experiments, isotopically enriched samples of Li were used (96% ^6Li and 99.99% ^7Li). This ensured that Li-isotope shifts could be identified as clearly as possible and thus helped in distinguishing the matrix products that did contain lithium from those that did not. For the suggested experiment, however, natural lithium will suffice. Apart from the use of natural lithium, Protocol 8 follows the original experiments of Andrews fairly closely.[35] Note that molten lithium will attack copper and is therefore likely to leak from the Knudsen cell if a copper gasket is used; a tantalum gasket should remain intact.

Protocol 8.
The matrix reaction of CCl_4 with Li atoms: the IR spectrum of CCl_3

Caution!   Carbon tetrachloride is toxic by inhalation and should be transferred from one vessel to another only in a fume cupboard.

Equipment

- Cold cell with a heater and temperature controller
- Knudsen cell with a tantalum gasket

- CsBr or CsI external windows for the cold cell (optional)
- IR spectrometer

Materials

- CCl$_4$ (of 99% purity or better) — **toxic**
- Li metal (in oil) — **reacts violently with water**
- High purity argon
- Petroleum ether, b.p. 40–60 °C — **flammable**

Illustrates

- Matrix IR spectroscopy
- Matrix reactions with metal atoms
- Annealing matrices

1. First prepare the source of Li atoms, as follows:

   (a) Cut a piece of rubber (e.g. from a rubber bung) to form a small plug which can fit snugly into the orifice of the Knudsen furnace and seal it (cf. Fig. 3.6).

   (b) Cut a small piece of lithium metal under oil (a cube measuring about 5 mm on each edge is about right), wash it free of oil with petroleum ether, dry it quickly by dabbing on a dry filter paper and by evaporation, and transfer it to the Knudsen cell.

   (c) Flush the cell with argon (this need not be high purity gas), and maintain a flow of the gas while quickly screwing on the threaded plug, using a tantalum gasket.

   (d) With argon still flowing, seal the orifice of the cell with the small rubber plug.

   NB The whole operation of cutting the lithium and transferring it to the Knudsen cell is best carried out under an argon atmosphere in a glove box or glove bag, if one is available.

   (e) Transfer the Knudsen cell containing the lithium to the cold cell, and assemble the furnace, not forgetting to place the thermocouple in the correct location (see Fig. 3.7).

   (f) Slightly loosen the rubber plug to ensure that it comes out of the orifice when vacuum is applied to the cold cell but still maintains the seal. Some practice may be needed to get this right every time.

   (g) Complete the assembly of the cold cell, with the Knudsen cell and furnace installed, and then evacuate it using the backing–roughing pump. Check that the rubber plug comes out of the orifice of the Knudsen cell during the evacuation and falls to the bottom of the shroud. If the plug remains in place, readmit air to the shroud and loosen it a little more.

2. At this stage, the cold cell can be left under vacuum at room temperature until the gas bulb is mounted on the spray-on line, or it can be cooled while the mixture of argon and CCl$_4$ is being prepared.

3. Set up a preparative vacuum line (cf. Fig. 4.1) with a 1 litre gas bulb and a cylinder of Ar.

Protocol 8. *Continued*

4. Place about 5 ml of carbon tetrachloride in a sample ampoule and degas it as described in Protocol 2, Chapter 4.

5. Prepare a gas mixture of Ar and CCl_4 with a matrix ratio (Ar:CCl_4) of 200:1, as described in Protocol 3, Chapter 4.

6. Transfer the bulb containing the gas mixture to the spray-on line of the cold cell.

7. Following Protocol 7, Chapter 4, begin depositing the gas mixture as a matrix at 15 K by the slow deposition technique. Adjust the spray-on rate to about 2 mmol h^{-1}.

    NB The original paper gave 15 K as the temperature for deposition, but deposition at the base temperature of the cold cell (e.g. 10–12 K) may work equally well.

8. Monitor the spray on by recording IR spectra.

9. Heat the Knudsen furnace to about 700 K.

10. Rotate the cold window so that the beam of metal atoms is deposited more or less perpendicularly on to it.

11. Record IR spectra at intervals. There is no need to halt deposition while recording the spectra, although the cold window will probably need to be rotated for this purpose.

12. Look for the strong CCl_3 absorption at 898 cm^{-1} and the LiCl absorption at 580 cm^{-1} (cf. Fig. 6.12). Increase the furnace temperature, if necessary, to increase the rate of Li effusion.

13. When a sufficiently intense CCl_3 band at 898 cm^{-1} can be seen, switch off the Knudsen furnace, wait a few minutes for it to cool, and then stop the flow of matrix gas.

14. Record an IR spectrum.

15. Anneal the matrix briefly at 35 K, following instruction 5 of Protocol 1, and record another IR spectrum.

16. Further annealing may be carried out if desired. Record an IR spectrum after each cycle.

The interpretation of the product IR spectra in the original experiments required a great deal of work. Isotope shifts due to [6]Li-, [7]Li- and ^{13}C-substitution were all utilized, together with careful comparison of the observed IR bands with the known matrix spectra of stable products, such as C_2Cl_6.[35] Figure 6.12 shows spectra presented in the original paper. The product spectra obtained in this experiment should closely resemble Fig. 6.12(c), although if KBr external windows have been used, the spectra will be cut off below 400 cm^{-1}. In Fig. 6.12, the very strong bands marked P are due to CCl_4. The

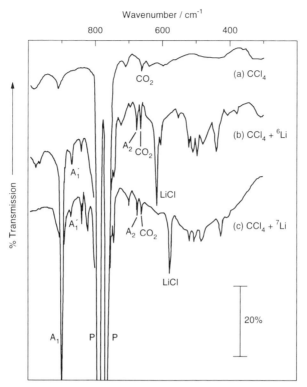

Fig. 6.12 Matrix IR spectra showing the formation of CCl_3 from the reaction of Li with CCl_4. (a) CCl_4 in Ar at 15 K, with Ar:CCl_4 = 200:1; (b) the same mixture co-deposited with [6]Li; (c) co-deposited with [7]Li. Spectra (b) and (c) are offset for clarity. Band assignments: P, CCl_4; A_1 and A_2, CCl_3, A_1', [13]CCl_3. Bands due to LiCl and CO_2 are also identified. *(Adapted with permission from ref. 35. © 1968. American Institute of Physics.)*

bands marked LiCl belong to lithium chloride, and show a significant Li-isotope shift, as expected. In contrast, the bands marked A_1 and A_2 show little or no Li-isotope shift, and are assigned to C–Cl stretches of the trichloro-methyl radical. The small satellite band marked A_1' is due to the 1% of naturally occurring [13]CCl_3, as was confirmed in experiments with ^{13}C-labelled CCl_4. Close inspection of the bands may also reveal splittings due to [35]Cl- and [37]Cl-isotopomers.

In the annealing process, the bands of CCl_3 should diminish in intensity, while bands due to C_2Cl_6 should increase. In Ar matrices, C_2Cl_6 has IR bands at 792, 786, and 684 cm^{-1}, but the first two will be obscured by intense CCl_4 absorptions; so only the last will be seen to grow.

In the procedure described for this experiment, no method for determining the rate of Li-atom effusion has been suggested, because it is not really necessary. It should be noted, however, that an estimate of this rate can be obtained, if desired, by a titrimetric technique. The cold window is replaced with a glass or quartz plate, then lithium from the Knudsen cell is deposited on it under full vacuum, but at room temperature, for an appropriate time. The deposited lithium is finally dissolved in water and titrated quantitatively with standard acid.[36] Alternatively, a quartz microbalance could be used (see Chapter 4, Section 3.3.1).

2.1.6 IR detection of xenon dichloride

Xenon dichloride was first generated in matrices by passing a mixture of xenon and chlorine through a microwave discharge.[37] A broad, structured IR absorption was observed at 313 cm^{-1} (Fig. 6.13(b)) and this was assigned to the v_3 mode (asymmetric stretch) of XeCl$_2$. Naturally occurring Xe has seven isotopes above 1% abundance. With seven Xe and two Cl isotopes, XeCl$_2$ can be expected to have IR absorptions with complex structures. Confidence in the assignment of the 313 cm^{-1} feature to XeCl$_2$ was greatly strengthened by the close match between the shape of the experimentally observed absorption (Fig. 6.13(b)) and a calculated band shape based on the natural abundances of all the isotopes and assuming a linear structure for the dichloride (Fig. 6.13(a)).

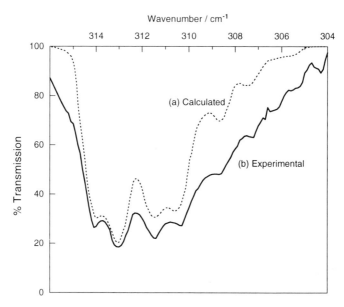

Fig. 6.13 The v_3 band (asymmetric stretch) of matrix isolated XeCl$_2$. (a) the spectrum calculated from natural abundances of the Xe and Cl isotopes; (b) the experimentally recorded spectrum. The matrix was formed by passing a mixture of Xe and Cl$_2$ through a microwave discharge followed by deposition at 20 K. The calculated absorption profile was based on the assumption of a linear structure for XeCl$_2$ and a Gaussian shape for each component, with a σ value of 0.37 cm^{-1}. *(Adapted with permission from ref. 37. © 1967, American Chemical Society.)*

Subsequently, $XeCl_2$ was generated by photolysis of Cl_2 in Xe matrices[38] and was also observed by Raman spectroscopy.[38,39] The Raman studies revealed a vibrational band of $XeCl_2$ at 254 cm^{-1}, which was identified as belonging to the ν_1 mode (symmetric stretch). These studies were supplemented, and largely confirmed, in matrix experiments with isotopically enriched xenon and chlorine.[40]

The observation of the IR band of matrix-isolated $XeCl_2$ is a reasonably easy exercise, which can be accomplished either by microwave generation of the dichloride or by matrix photolysis. Protocol 9 gives a procedure for the former approach. To permit observation of the 313 cm^{-1} absorption of $XeCl_2$, the cold cell should be fitted with CsBr or CsI outer windows, and the IR spectrometer should have a wavenumber range extending down to 300 cm^{-1} or below. Chlorine of 99.5% purity can be obtained in small cylinders, and this should be sufficiently pure for the experiment. If preferred, it can, with care, be purified and degassed on a vacuum line by trap-to-trap distillation.

Protocol 9.
The generation of XeCl₂ by microwave discharge

Caution! Chlorine is toxic and should not be allowed to escape from the cylinder into the open laboratory. Store the cylinder in a fume cupboard or other well ventilated place when it is not in use, and vent any small amounts of excess Cl_2 into a fume cupboard or to the outside of the building. Stainless steel or Monell metal valves are recommended.

Equipment
- Cold cell with CsBr or CsI external windows
- Microwave generator
- IR spectrometer with wavenumber range extending down to 300 cm^{-1} or lower

Materials
- Cl_2 (of 99.5% purity or better)   **toxic, irritant**
- High purity xenon

Illustrates
- Matrix IR spectroscopy
- Microwave generation of reactive species

1. Set up the cold cell for IR spectroscopy and microwave generation, with a Pyrex inlet tube of about 15 mm o.d., and the cavity of the microwave umit surrounding the tube about 10 cm away from the shroud. Compare Fig. 6.8, which shows a similar cavity and tube arrangement with an ESR shroud.

2. Prepare a mixture of Xe and Cl_2 with a matrix ratio (Xe:Cl_2) of between 100:1 and 200:1, following Protocol 1 (this chapter) and Protocol 3, Chapter 4.

3. Transfer the gas bulb to the spray-on line of the cold cell.

Protocol 8. *Continued*

4. Following Protocol 7, Chapter 4, begin depositing the gas mixture as a matrix at 10–20 K by the slow deposition method. The rate of deposition is not critical; 1–5 mmol h^{-1} should be about right.

5. Once a steady rate of deposition has been established, switch on the microwave generator and adjust the power and tune the cavity to give a full rich glow in the Pyrex inlet tube.

6. Continue depositing the gas mixture through the microwave discharge for about 20–30 min, then record an IR spectrum. It would be best to switch off the microwave generator while the spectrum is recorded, but the gas flow can be left unaltered.

7. Continue depositing the gas mixture through the microwave discharge until a reasonable intensity of the 313 cm^{-1} absorption of XeCl$_2$ has built up.

8. Switch off the microwave generator and then stop the gas flow.

9. Record a final IR spectrum.

The matrix IR spectrum recorded at the end of this experiment should closely resemble Fig. 6.13(b). The original papers can be consulted for more details.[37–40]

If a microwave generator is not available, matrix-isolated XeCl$_2$ can be produced by photolysis of Cl$_2$ in a Xe matrix. This was first described by Howard and Andrews,[38] who used matrix ratios (Xe:Cl$_2$) of about 50:1 to 200:1, and carried out photolysis with either an argon-ion laser or a high-pressure Hg arc. The latter should be equipped with a water filter (cf. Fig. 3.3) and should be additionally filtered to allow through light in the range 400–550 nm (i.e. visible light at the blue end of the spectrum).

2.1.7 Matrix isolated POCl

The reaction of phosphoryl chloride (POCl$_3$) with silver at 1000–1200 K produces POCl, the phosphorus analogue of nitrosyl chloride. The reaction can be performed in a heated reaction tube (see Chapter 3, Section 2.2.4), and the product co-deposited with an excess of argon as a matrix.[41] In the original work, POCl$_3$ vapour was passed over silver in a quartz tube at about 1100 K. The pyrolysate was then mixed, in the gas phase, with an excess of argon about 5 cm in front of the cold surface on which the matrix was deposited. In these experiments, the matrix was deposited on a polished copper block instead of a transparent window and IR spectra were recorded by reflection.

Phoshoryl chloride is readily and cheaply available (some grades have a purity of 99.999%); so this would be an easy experiment to carry out to test a reaction tube, without the need for very high temperatures. Figure 6.14 shows an IR spectrum of matrix-isolated POCl in which all three fundamental

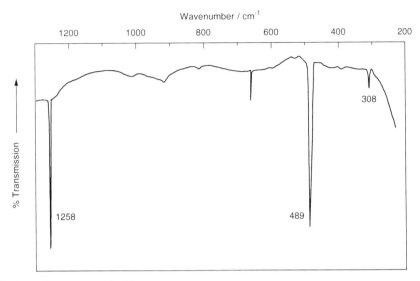

Fig. 6.14 IR spectrum of POCl in an argon matrix. The band assignments are: ν(PO) 1258, ν(PCl) 489, δ(OPCl) 308 cm^{-1}. *(Adapted with permission from ref. 41. © 1983, Johann Ambrosius Barth, Hüthig GmbH.)*

vibrations can be identified. The assignments were originally made with the help of ^{18}O-substitution and a force-field calculation. Note that the bending mode absorption at 308 cm^{-1} will be observable only if CsBr or CsI outer windows, rather than KBr, are fitted to the shroud.

2.1.8 Silicon dihalides

The silicon dihalides, SiF_2, $SiCl_2$, and $SiBr_2$, have all been isolated in matrices following passage of the corresponding tetrahalides over elemental silicon at 1200–1400 K and the resulting IR spectra have been reported.[42,43] The silicon tetrahalides are readily available at reasonable cost. The tetrachloride, in particular, is a liquid and therefore convenient to handle, and can be obtained as semiconductor grade with 99.999% purity. The generation and matrix isolation of $SiCl_2$, or either of the other dihalides, would seem, therefore, to offer an interesting and not too difficult test for a reactive tube set-up. The reader should consult the original papers for experimental details.

2.2 Larger organic species in matrices

Cyclobutadiene is often regarded as the first example of an unstable species to be characterized in low-temperature matrices that is also a 'real organic molecule.' The study of this species has been discussed briefly in Chapter 1 (Section 1.3). Despite this historically pivotal role, however, the matrix isolation of cyclobutadiene has not been included as a suggested experiment in this chapter for two reasons. First, the preparation of the photoprecursor, α-pyrone, is somewhat inconvenient, and second, the generation of cyclo-butadiene from this precursor requires many hours of photolysis. The ex-amples chosen should prove less time consuming, though in most cases the starting materials are too unstable to be available commercially and thus have to be synthesized.

2.2.1 The matrix photolysis of phenyl azide

Photolytic or thermal decomposition of phenyl azide (PhN$_3$) results in loss of N$_2$, and generates the very reactive species, phenylnitrene. In normal solution-phase reactions, the nitrene reacts further to give a variety of products, depending on the other species present. The photolysis of phenyl azide in an argon matrix was first reported by Chapman and Le Roux.[44] The results, at the time, were something of a surprise. Before the matrix experiments had been carried out, theoretical studies had suggested that the most stable singlet product from the unimolecular decomposition of PhN$_3$ would be a bicyclic azirine, rather than either the singlet nitrene or an azacycloheptatetraene (Scheme 6.3). In contradiction of these predictions, the matrix photoproduct from phenyl azide had an intense band at 1895 cm^{-1}, and this could only be explained if the product was 1-aza-1,2,4,6-cycloheptatetraene (Scheme 6.3), which has a strained ketenimine structure. The assignment of the 1895 cm^{-1} band to ν(C=C=N) of the strained ketenimine was later confirmed by ^{15}N labelling, and the ring expansion of phenylnitrenes was found to be quite general for a range of 3- and 4-substituted analogues.[45] With carefully chosen photolysis wavelengths, it is possible to isolate the triplet phenylnitrene in matrices.[46] The triplet can absorb a second photon to regain the singlet manifold and give the azacycloheptatetraene; so in normal matrix photolysis with broad-band light, the triplet is not observed or is observed only weakly.

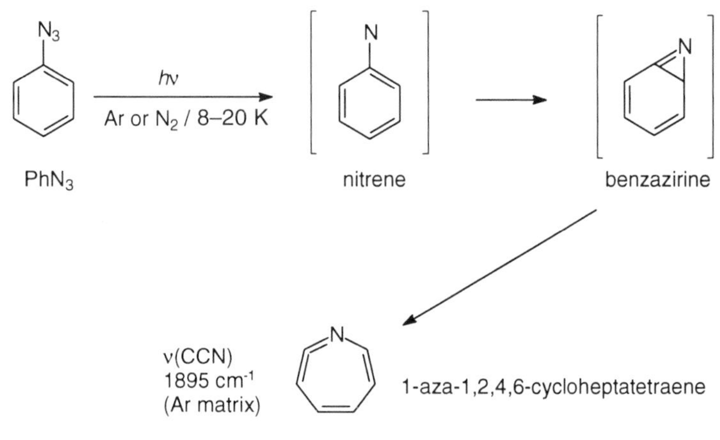

Scheme 6.3

Unravelling the mechanism of the photolysis of phenyl azide has proved unexpectedly complicated; the whole story has been reviewed recently.[47]

Phenyl azide can be prepared conveniently by diazotization of phenyl-hydrazine.[48] It is stable enough to be stored for short periods at room temperature, provided it is protected from exposure to light. It keeps indefinitely in a freezer. Phenyl azide is not one of the most hazardous of azides, but it can explode if heated strongly. The procedure for its preparation must be followed with care. The purification of the azide requires first a steam distillation, and then a final distillation under reduced pressure (b.p. 49–50°C at 5 mmHg (6.6 mbar)). The heating bath for the final distillation should be kept below 80°C at all times. In many years of working with this compound, the author's group has never experienced an explosion. For matrix work, store the azide in ampoules (cf. Fig. 4.1) containing no more than 2–5 ml.

The matrix photolysis of PhN_3 can be followed by IR and UV–visible spectroscopy. The UV–visible absorptions of the azacycloheptatetraene, however, are rather weak; so an IR experiment seems best as an introduction to this area of matrix chemistry. Phenyl azide has a vapour pressure of about 1–1.5 mbar at room temperature. This is just enough to allow it to be incorporated into a premixed matrix gas, without the need for special gauges. Use of a sensitive gauge, such as an oil-filled manometer (Fig. 3.3), is advised, if reasonably accurate matrix ratios are desired. Protocol 10 gives the procedure for a typical matrix IR experiment with PhN_3.

Protocol 10.
Photolysis of phenyl azide in an Ar or N$_2$ matrix

Caution! Phenyl azide is explosive when heated strongly. Note also that striking an arc lamp too close to sensitive electronic equipment can result in malfunction or damage (see Chapter 3, Section 2.1.1, subsection *i*).

Equipment
- IR spectrometer
- Calibrated expansion bulb (optional)

- Medium- or high-pressure Hg arc with water filter

Materials
- Phenyl azide   **potentially explosive**
- High purity Ar or N$_2$

Illustrates
- Matrix IR spectroscopy

- Matrix photolysis

1. Set up a preparative vacuum line (cf. Figs 4.1 and 4.4) with a 1 litre gas bulb and a cylinder of Ar or N$_2$. A calibrated expansion bulb can be incorporated if desired.

Protocol 10. *Continued*

2. Place 1–2 ml of the phenyl azide in a sample ampoule, and degas it using the procedure of Protocol 2, Chapter 4. Observe the precautions for handling potentially explosive compounds.

3. Take a 1 litre gas bulb, cover it with aluminium foil to exclude light, and, following Protocol 3 or 6 in Chapter 4, fill it to 500–800 mbar with a mixture of Ar or N_2 and PhN_3, with a matrix ratio (host:PhN_3) of between 500:1 and 1000:1. Note the following:

   (a) Phenyl azide has a vapour pressure at room temperature of about 1–1.5 mbar; on no account attempt to increase this by warming the sample ampoule, as this will result in condensation of the azide in cooler parts of the vacuum line.

   (b) To prepare the gas mixture first introduce 0.2–0.8 mbar of the azide vapour into the gas bulb. This may be achieved in one of two ways:

      i. Follow Protocol 3, Chapter 4, and fill the bulb directly with 0.2–0.8 mbar of the azide vapour, using a sensitive pressure gauge to measure the pressure as accurately as possible.

      ii. Follow Protocol 6, Chapter 4, use a calibrated expansion bulb, and expand 1 mbar of the azide vapour into the gas bulb several times, until it is calculated that the required pressure has been attained. For example, 1 mbar of azide vapour expanded ×10 into the gas bulb gives a final pressure of 0.1 mbar; so repeating this operation five times fills the bulb to 0.5 mbar.

   (c) Complete the preparation of the matrix-gas mixture by filling the bulb to the appropriate pressure with the selected host gas (cf. Protocols 3 and 6, Chapter 4).

4. Transfer the bulb containing the gas mixture to the spray-on line of the cold cell.

5. Following Protocol 7, Chapter 4, deposit the gas mixture at 10–20 K.

6. Monitor the progress of matrix deposition by recording IR spectra at intervals.

7. When the very strong v(NNN) bands of the azide near 2100 cm^{-1} have reached about 10% transmission (absorbance = 1) (cf. Fig. 6.15(a)), stop the deposition and record another IR spectrum.

8. Irradiate the matrix with a medium- or high-pressure Hg arc through a water filter (to remove IR radiation). No other light filtration is necessary.

9. The time needed to photolyse most of the azide will vary with the lamp power, but 30–180 minutes would be a typical range. Record an IR spectrum after 10–15 min, and monitor the progress of the photolysis by recording IR spectra at intervals thereafter.

10. On finishing the experiment, it may be found that the vacuum lines take several hours to regain their usual low pressures; this is quite normal. Phenyl azide appears to be adsorbed strongly on both glass and metal interior surfaces. If the time taken to free the system of residual phenyl azide becomes excessive, assist the process by gently heating affected parts of the lines with warm air.

Figure 6.15 shows IR spectra recorded before and after photolysis of phenyl azide in a nitrogen matrix. The very strong absorption belonging to the asymmetric NNN stretching mode of PhN_3 is split by Fermi resonance and appears as a series of bands near 2100 cm^{-1} (Fig. 6.15(a)). Most of the azide was removed in the photolysis, while the $\nu(CCN)$ band of the azacyclo-heptatetraene product at 1891 cm^{-1} arose (Fig. 6.15(b)). Table 6.4 lists the IR bands that have been assigned to the azacycloheptatetraene with reasonable confidence, and these should all be visible in the photoproduct spectrum in this

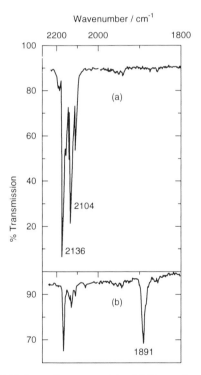

Fig. 6.15 Photolysis of PhN_3 in a nitrogen matrix. (a) IR spectrum of PhN_3 in N_2 at 12 K, N_2:PhN_3 = 1000:1; (b) the same matrix after 2 hours photolysis. Note the change of wavenumber scale at 2000 cm^{-1}. The very strong, split absorption near 2100 cm^{-1} in (a) belongs to the $\nu(NNN)_{as}$ mode of PhN_3; the 1891 cm^{-1} product band in (b) belongs to the $\nu(C=C=N)$ mode of 1-aza-1,2,4,6-cycloheptatetraene (Scheme 6.3).

Table 6.4. IR bands of 1-aza-1,2,4,6-cycloheptatetraene (Scheme 6.3) in N_2 at 12 K.[a]

Wavenumber / cm^{-1}	Relative intensity	Assignment
3024	0.05	ν(C–H)
1891	1.0	ν(C=C=N)
1346	0.15	
1110	0.1	ν(C–N)
978	variable[b]	
691	0.35	
664	0.25	
655	0.1	
585	0.15	
516	0.1	

[a] Data from ref. 45.
[b] This band has variable relative intensity due to overlap with a band of a secondary photoproduct.

experiment. The band positions and intensities do not vary significantly from Ar to N_2 matrices.

If UV and IR polarizers are available, the matrix photolysis of PhN_3 can be carried out with plane-polarized light and dichroism in the IR spectrum of the residual azide measured (see Chapter 5). This has already been described in two papers,[49,50] the second of which also gives a full tabulation of the matrix IR bands of PhN_3. The observed pattern of IR dichroism in the photoselected PhN_3 confirmed that it has a planar or near-planar conformation in matrices, in accord with expectations from earlier solution studies and the crystal structure of a substituted phenyl azide.

2.2.2 Diazocyclopentadiene and the carbene, cyclopentadienylidene

Small carbenes, such as CCl_2, had been generated in matrices in the early years of development of the subject, in, for example, reactions between lithium atoms and tetrahalomethanes. One of the first examples of a larger carbene to be studied in matrices was cyclopentadienylidene (Cp:, Scheme 6.4).[51,52] This

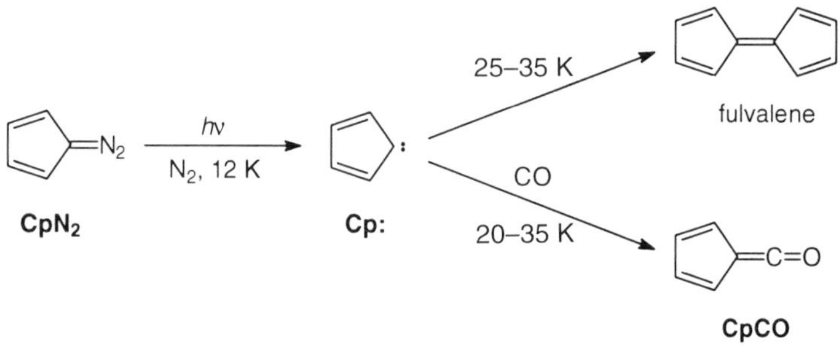

Scheme 6.4

carbene was generated by photolysis of diazocyclopentadiene (CpN_2) and was found to undergo thermal dimerization to yield fulvalene when the matrices were warmed. If CO was included in the matrix-gas mixture, thermal reaction with CO to give a ketene (CpCO) occurred on annealing.

Matrix-isolation studies of cyclopentadienylidene have been quite varied and have proved very rewarding over many years. Therefore some experiments with this compound can be recommended. Diazocyclopentadiene is most conveniently prepared from freshly distilled cyclopentadiene and *p*-toluenesulfonyl azide by the action of diethylamine or ethanolamine.[53] It can be purified by trap-to-trap distillation on a vacuum line, and should then be stored in glass ampoules in a freezer, or in Dewar vessels kept topped up with liquid nitrogen. It should be protected from the light at all times. In the author's experience, one such sample, stored in liquid N_2, remained in good condition for several years. The compound is a dark, orange or red liquid at room temperature and freezes to a yellow solid.

Diazocyclopentadiene must be treated with respect. Besides being undoubtedly toxic, it is potentially explosive. It is therefore recommended that only about 1–2 g of the material should be stored in each ampoule and that all safety precautions for handling potentially explosive materials should be adopted. Despite the hazards, the author has experimented with this compound for many years without mishaps. Diazocyclopentadiene has a vapour pressure of 5–10 mbar at room temperature, and therefore can be incorporated into premixed matrix gases without the need for specially sensitive pressure gauges.

The thermal dimerization of Cp: to give fulvalene and the reaction between Cp: and CO can both be followed in matrices by either UV or IR spectroscopy. The dimerization provides a spectacular product UV spectrum, however, which is well worth experiencing; so it is suggested that this reaction be studied in a UV experiment, and the reaction with CO in an IR experiment (see below). Some further suggestions for experiments with polarized light and O_2 are also made.

i. Thermal dimerization of the carbene Cp: to give fulvalene: matrix UV
  spectroscopy
In the first matrix experiments with CpN_2 it was found that this compound has a pronounced tendency to condense in the matrix in pairs of molecules or aggregates. This was apparent from UV spectra of the matrices after photolysis. The dimer, fulvalene, has a very characteristic UV spectrum, consisting of a series of sharp bands; so this product was very easy to pick out in the spectra. The UV absorption bands of fulvalene dominated the matrix photoproduct spectra in all but extemely dilute matrices. Figure 6.16 compares the results from photolysis of CpN_2 in N_2 matrices with matrix ratios ($N_2:CpN_2$) of 1000:1 and $1.5 \times 10^6:1$. The latter matrix ratio may still be a record for high matrix dilution, but even so ripples due to the sharp fulvalene bands can just be discerned on the broad absorption of the carbene (Fig. 6.16(c)). With more

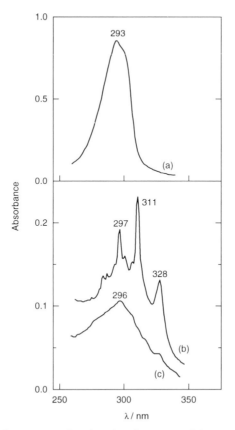

Fig. 6.16 UV absorption spectra showing the photolysis of CpN_2 in N_2 matrices at 20 K. (a) $N_2:CpN_2$ = 1000:1, before photolysis; (b) $N_2:CpN_2$ = 1000:1, after complete photolysis; (c) $N_2:CpN_2$ = 1.5×10^6:1, after complete photolysis. See the text for a discussion. *(Adapted with permission from ref. 52. © 1980, American Chemical Society.)*

concentrated matrices (Fig. 6.16(b)), the fulvalene spectrum can be seen clearly.

For an interesting UV experiment with CpN_2, prepare a mixture of CpN_2 and N_2 with a matrix ratio of between 20 000:1 and 50 000:1, following Protocol 6, Chapter 4 and Section 2.6.1 of Chapter 4. Deposit this gas mixture at 10–20 K until the UV absorption of CpN_2 has an absorbance of about 1.0 (cf. Fig. 6.16(a)). Photolyse with a medium- or high-pressure Hg arc through a water filter and follow the progress of photolysis by recording UV spectra at intervals. Complete photolysis of the CpN_2 will require only a few minutes, so record UV spectra at intervals of no more than a minute or two. Continue photolysing until all the CpN_2 has disappeared. The final spectrum should resemble that shown in Fig. 6.16(c), although the fulvalene bands will be more apparent. Anneal the matrix at 25 K for a few minutes (Instruction 5 of Protocol 1) and

Table 6.5. IR bands (cm^{-1}) of cyclopentadienylidene (Cp:, Scheme 6.4) in N$_2$ at 20 K.[a]

1345 w[b]	922 w
1335 m	703 s
1101 w	577 w
1074 w	

[a] Data from ref. 52.
[b] Qualitative band intensities: s strong, m medium, w weak.

record another UV spectrum. There should be distinct growth of the fulvalene bands, indicating dimerization of the carbene. Continue annealing at progressively higher temperatures (up to about 35 K), and record a UV spectrum after each annealing cycle. The fulvalene bands will grow further.

The same reaction can be investigated by IR spectroscopy, but very high matrix ratios would require very thick matrices for adequate intensity in the IR bands. For IR observation of Cp:, therefore, a matrix ratio (N$_2$:CpN$_2$) of about 2000:1 should be chosen. The original paper should be consulted for details of this experiment;[52] Table 6.5 list the IR bands which can be confidently assigned to the carbene. The most prominent band in the IR spectrum of Cp: is found at 703 cm^{-1}, and this clearly belongs to an out-of-plane C–H deformation mode.

ii. The thermal reaction between Cp: and CO: matrix IR spectroscopy
This experiment requires a premixed matrix gas, consisting of CpN$_2$ in a mixed host gas of 10% CO in N$_2$. A method of preparing this mixture is given in Protocol 11.

Once the gas mixture is made up according to the method of Protocol 11, it can be deposited as a matrix at 10–20 K, then photolysed exactly as described for phenyl azide in Protocol 10. Irradiate the matrix with UV light until all, or nearly all, the CpN$_2$ has been photolysed, then record an IR spectrum. This should show the presence of the carbene (Cp:); see Table 6.5 for the expected IR bands for this species, which will not differ greatly in position or relative intensity in CO and N$_2$ matrices.

Anneal the matrix briefly at 24–25 K (cf. instruction 5 of Protocol 1), and then record another IR spectrum. The thermal reaction between the carbene and CO should result in a distinct decrease in intensity of the IR bands of the carbene and growth of the IR bands of the ketene product, CpCO (Scheme 6.4). The IR absorption of carbon monoxide will be too intense to reveal any decrease in intensity. Figure 6.17 shows the result of annealing the matrix on the strong out-of-plane C–H modes of Cp: and CpCO. Table 6.6 lists the IR bands of CpCO which can be reliably assigned, and all but the weakest of these should be visible in this experiment.

Protocol 11.
Preparation of a matrix-gas mixture with a mixed host gas:
N_2:CO:CpN_2 = 9000:1000:1

The following is a general procedure for preparing premixed matrix-gas mixtures containing a guest species and a mixed host gas. The exact volumes and pressures specified are suggestions for the particular mixture of the experiment; it is obvious that there are alternative ways of arriving at the same volume ratios.

Caution!   Diazocyclopentadiene is toxic and has been known to explode during vacuum distillation. Do not heat this compound and observe all suggested precautions when making up the gas mixture.

Equipment
• Calibrated expansion bulb with a volume of about 10 cm^3
• 1 litre gas bulb

Materials
• Diazocyclopentadiene   **toxic, potentially explosive**
• High purity N_2
• High purity CO (cylinder or glass bulb)   **toxic, flammable**

1. Set up a preparative vacuum line (cf. Fig. 4.4) with a 1 litre gas bulb, the calibrated expansion bulb (about 10 cm^3), and a cylinder of N_2.

2. Connect the source of CO to the vacuum line, either as a second cylinder or as a gas bulb on a vacant port. (See Chapter 4, Section 2.4.4 for advice on connecting two cylinders.)

3. Condense 1–2 ml of the diazocylopentadiene (CpN_2) into a sample ampoule by trap-to-trap distillation and degas it using the procedure of Protocol 2, Chapter 4. Observe the precautions for handling potentially explosive compounds.

4. Prepare the mixture of N_2, CO, and CpN_2 as follows (cf. Protocol 6, Chapter 4):

   (a) Take a 1 litre gas bulb, cover it with aluminium foil to exclude light, and fill it to 0.05 mbar with CpN_2, by expanding 5 mbar of the vapour of the diazo compound ×100 from the expansion bulb into the 1-litre bulb.

   (b) Dilute the CpN_2 with CO to a final pressure of 50 mbar. This gives a gas mixture with CO:CpN_2 = 1000:1.

   (c) Dilute the mixture of CO and CpN_2 with N_2 to a final pressure of 500 mbar

5. The gas mixture now has a matrix ratio, N_2:CO:CpN_2 = 9000:1000:1 and is ready to be transferred to the spray-on line of the cold cell.

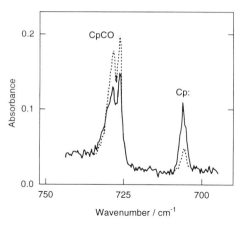

Fig. 6.17 IR spectra showing the reaction of the carbene Cp: with CO in a nitrogen matrix (cf. Scheme 6.5). The solid line shows the δ(CH) bands of CpCO and Cp:, generated by photolysis at 12 K of CpN_2 in an N_2 matrix containing 10% CO, N_2:CO:CpN_2 = 9000:1000:1. The broken line shows the spectrum of the same sample after annealing at 24 K. *(Adapted with permission from ref. 52. © 1980, American Chemical Society.)*

Table 6.6. IR bands (cm^{-1}) of the ketene CpCO (Scheme 6.4) in CO matrices.[a]

2133 vs[b]	1326 w
2130 s	1080 w
2082 w	927 w
2079 w	901 w
1444 w	737 m
1432 m	699 w
1396 w	640 w
1385 w	582 m
1380 w	527 w

[a] Data from ref. 52.
[b] Qualitative band intensities: vs very strong, s strong, m medium, w weak.

The thermal reaction between Cp: and CO observed in this experiment is quite remarkable, because it seems to be spin-forbidden and yet clearly has a very low activation energy. The carbene Cp: almost certainly has a triplet ground state—and indeed reacts with molecular oxygen, which has a triplet ground state, even more readily than with CO. This aspect of carbene reactivity has yet to be fully explored.

iii. Photolysis of diazocyclopentadiene with polarized light
The diazocyclopentadiene molecule has a rigid C_{2v} structure, which makes it ideal for photoselection studies. Polarized photolysis of CpN_2 has been carried

out in both N_2 and CO matrices and relatively high degrees of polarization were achieved.[52] These experiments are discussed in Chapter 5 (see Sections 1.5 and 2.4.2) and would provide some good exercises in matrix photolysis with polarized light.

iv. The reaction of Cp: with O_2 to give a carbonyl oxide

For those seeking a real challenge in matrix technique, a study of the reaction between the carbene Cp: and O_2 can be recommended. The carbene reacts even more readily with O_2 than with CO, yielding a carbonyl oxide (Scheme 6.5).[54-56] The carbonyl oxide can be observed by both UV–visible and IR spectroscopy, but it is extremely photolabile. The experiments have to be carried out in a completely blacked-out laboratory, therefore, or in carefully covered equipment.

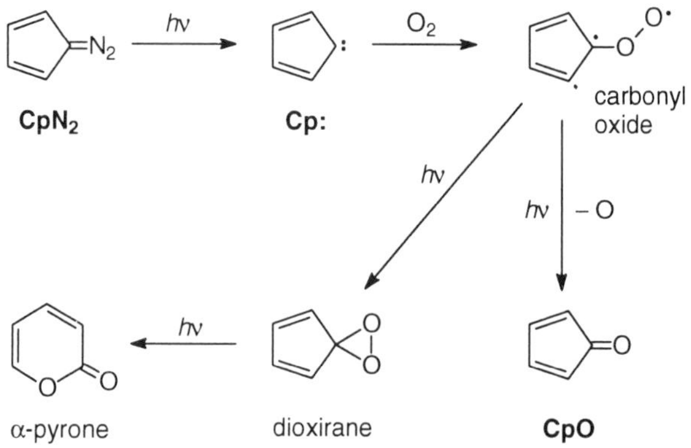

Scheme 6.5

The secondary photoproducts from the carbonyl oxide are cyclopentadienone (CpO), arising by expulsion of an O atom, and α-pyrone. The latter product is formed via a dioxirane intermediate, which has also been observed by matrix IR spectroscopy.[56] The end-on bonding of oxygen in the carbonyl oxide and sideways bonding in the dioxirane were established by isotopic labelling with $^{16}O^{18}O$.

With the necessary precautions to exclude room light from the cold cell, the carbonyl oxide can be generated in N_2 or Ar matrices containing 1–10% of O_2, in much the same way that the ketene CpCO is generated in matrices containing CO (see subsection *ii* above). A number of other carbonyl oxides can be studied in the same way. For further details on carbonyl oxides the reader should consult the original papers[54-56] and a review article.[57]

2.2.3 Tetrachlorocyclopentadienylidene

The tetrachloro analogue of diazocyclopentadiene is safer and easier to work with than the parent compound. It can be prepared conveniently from the commercially available hexachlorocyclopentadiene, by treatment with hydrazine hydrate, followed by oxidation with lead tetraacetate (Scheme 6.6).[58] It is a crystalline solid (m.p. 108–109°C) and can be purified by recrystallization from hexane. It is less volatile than diazocyclopentadiene and is therefore incorporated into matrices by sublimation from a side-arm (cf. Fig. 4.6), as described in Chapter 4, Section 3.3.

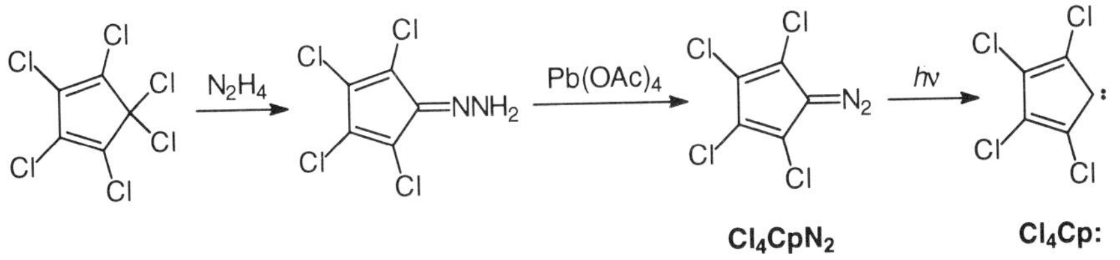

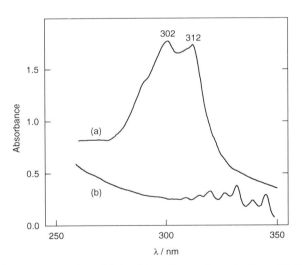

Scheme 6.6

The matrix photolysis of tetrachlorodiazocyclopentadiene (Cl_4CpN_2) produces the corresponding carbene Cl_4Cp: (Scheme 6.6). This carbene does not appear to dimerize in matrix-isolation conditions, presumably owing to the steric demands of the chlorine atoms next to the carbene centre, but it reacts with carbon monoxide[59] and oxygen[55,56] to give a ketene and a carbonyl oxide, respectively. The carbene has an interesting UV spectrum (Fig. 6.18(b)), showing extended vibrational coupling. It can also be observed by IR spectroscopy.

Fig. 6.18 UV absorption spectra of tetrachlorodiazocyclopentadiene (Cl_4CpN_2, Scheme 6.6) in a nitrogen matrix at 12 K. (a) Before photolysis; (b) after complete photolysis with $\lambda = 312 \pm 5$ nm. The product spectrum in (b) belongs to the corresponding carbene (Cl_4Cp:); see Table 6.7 for wavelength data. The matrix ratio was not determined. *(Adapted with permission from ref. 59. © 1985, Royal Society of Chemistry.)*

Table 6.7. UV and IR absorptions of the carbene, tetrachloro-cyclopentadienylidene (Cl_4Cp:), in N_2 matrices.[a]

UV: λ_{max} / nm	IR: $\bar{\nu}_{max}$ / cm^{-1}
304	1512 w[b]
310	1508 w
316	1356 w
320	1138 w
327	1127 m
332	892 w
340	882 w
345	703 w
	550 w
	530 w

[a] Data from ref. 59.
[b] Qualitative band intensities: m medium, w weak.

The matrix photolysis of Cl_4CpN_2 provides a good exercise in forming matrices by sublimation from a side-arm with simultaneous deposition of the host gas, generating a reactive intermediate photolytically, and observing it by UV or IR spectroscopy. Figure 6.18 shows UV spectra of the starting material and the carbene in an N_2 matrix; and Table 6.7 lists both the UV and the IR absorptions of the carbene.

2.2.4 Cyclopentadienone

In normal conditions, cyclopentadienone (CpO) readily dimerizes by a Diels–Alder reaction, and therefore cannot be isolated as a pure compound. It is stable in rigid low-temperature matrices, however, and is formed as a product in a variety of organic matrix reactions, such as the photo-oxidation of diazocyclopentadiene (Scheme 6.5). In argon matrices, CpO dimerizes on annealing at 38 K. Despite earlier reports of the observation of matrix isolated CpO, it was not until a thorough study of both photolytic and pyrolytic routes to this molecule was made by Maier *et al.*[60] that complete spectroscopic data for the matrix isolated species (UV and IR) were published.

The most convenient precursor for a matrix study of cyclopentadienone is probably *o*-benzoquinone, which can be prepared by oxidation of catechol (1,2-dihydroxybenzene) and purified by vacuum sublimation.[60] Interestingly, photolysis of the quinone in Ar matrices at long wavelengths ($\lambda > 300$ nm) results in ring opening to a bisketene (Scheme 6.7), while photolysis at 254 nm gives cyclopentadienone. To obtain CpO, it is important to use a low-pressure Hg arc, therefore. The original paper quotes ten hours as the time needed for complete photolysis. As an alternative, CpO can be generated from *o*-benzoquinone by pyrolysis at 1000 K. Matrix-isolated CpO has IR bands at 1727, 1724, 1332, 1136, 822, 632 and 458 cm^{-1}.

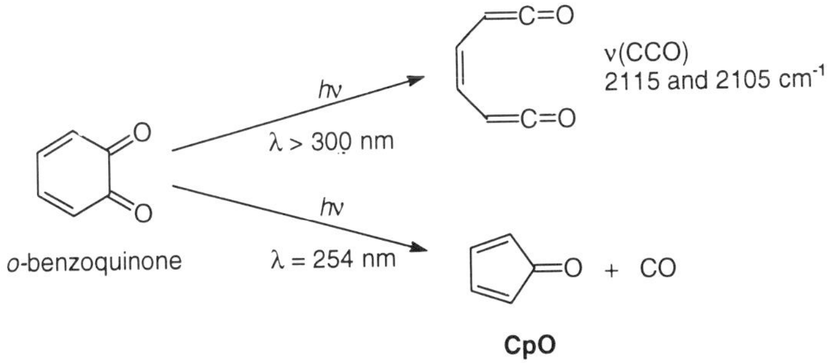

ν(CCO)
2115 and 2105 cm^{-1}

o-benzoquinone

$h\nu$
$\lambda > 300$ nm

$h\nu$
$\lambda = 254$ nm

CpO

+ CO

Scheme 6.7

2.2.5 The benzyl radical from pyrolysis of benzyl bromide

A straightforward pyrolysis experiment, using a precursor which can be bought cheaply, is the pyrolysis of benzyl bromide. In an early example of the combination of flash vacuum pyrolysis with matrix isolation, Angell *et al.*[61] found that pyrolysis of PhCH$_2$Br at about 1300 K produced the benzyl radical, which could be trapped in matrices. These workers used a liquid helium cryostat and were thus able to deposit neon matrices. They reported the IR, UV–visible absorption, fluorescence, and ESR spectra of the radical and emphasized one of the significant advantages of matrix isolation: that different types of spectra can be obtained for a reactive species under essentially identical conditions.

2.2.6 Pyrolysis and photolysis of 2-iodoacetic acid: matrix ESR spectroscopy

Other matrix experiments involving a readily available precursor are the pyrolysis and photolysis of 2-iodoacetic acid, which were first reported by Kasai and McLeod.[62] The pyrolysates were trapped in argon matrices and ESR spectra of the resulting radicals were obtained. Following pyrolysis at 770 K, only the methyl radical could be detected, but with pyrolysis at 570 K, the ESR spectrum of $\cdot$CH$_2$CO$_2$H was observed. UV photolysis of 2-iodoacetic acid in Ar matrices produced methyl radicals, formyl radicals and hydrogen atoms.

The pyrolysis experiments require only modest pyrolysis temperatures, and so could offer a straightforward way of testing a matrix ESR set-up. In the original work, 2-iodoacetamide was similarly studied and use was made of deuterium substitution.

2.2.7 UV–visible absorption spectra of the benzyl and tropylium cations in argon matrices

Cations can be studied in the gas phase by mass-spectrometry, but little information about the structures of the cations can be derived from this technique. The advantage of matrix isolation is that cations can be trapped at low temperatures and UV–visible and IR spectra obtained. One method of generating cations in matrices is by use of an argon resonance lamp (see Chapter 3, Section 2.4.1). Argon is passed through a microwave discharge then mixed with the matrix guest and finally deposited on the cold window. In the process, a proportion of the guest molecules becomes ionized. As explained in Chapter 3, it is not entirely clear by what process the ionization occurs. One view is that the excited argon atoms emit resonance radiation in the vacuum-UV and absorption of the high-energy photons ionizes the guest. Another view is that guest molecules collide with excited Ar atoms generated in the discharge and undergo energy transfer, followed by ionization. Whatever the mechanism of ionization, microwave discharges in Ar provide an efficient way of producing cations in matrices. To stabilize the cations, it is usual to include an electron trap, such as CCl_4, along with the guest in the matrix-gas mixture.

A simple experiment to demonstrate cation formation is the co-condensation of benzyl bromide vapour with argon from a discharge lamp, as described by Andrews and Keelan.[63] UV–visible spectra of both the benzyl cation and the tropylium cation can be observed. The experimental set-up is shown in Fig 3.8. Benzyl bromide has a vapour pressure of about 1 mbar at room temperature. It was deposited neat through a needle valve, while argon flowed through the quartz discharge tube (3 mm orifice) at about 0.5 mmol h^{-1}. If it is intended to carry out a similar experiment, begin by co-depositing benzyl bromide and argon for about 1 hour without the discharge, so as to observe the absorption of the starting material. Then switch on the microwave discharge and continue deposition for another two hours. Figure 6.19(a) shows a UV-visible absorption spectrum recorded after this procedure had been followed. Besides the absorption of benzyl bromide at short wavelengths, the spectrum has a broad maximum at 353 nm, which belongs to the benzyl cation ($PhCH_2^+$), and a shoulder at 263 nm, which belongs to the tropylium cation. Photolysis of the sample with a Pyrex-filtered mercury arc ($\lambda > 290$ nm) resulted in a decrease in intensity of the benzyl cation absorption and growth of that due to tropylium cation (Fig. 6.19(b)), showing that a photo-rearrangement of the benzyl cation had occurred (Scheme 6.8). Note that in these experiments, there was no need to add an electron trap, since bromine was present in the precursor, and bromide ion was presumably formed as the counterion. The reader should consult the original paper for full experimental details and a discussion of the results.[63]

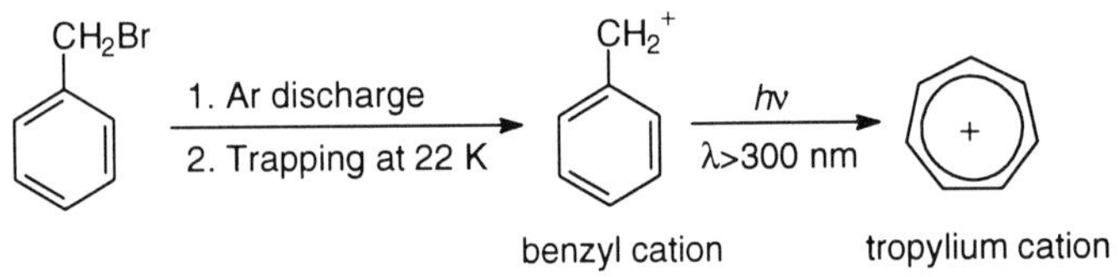

CH₂Br 1. Ar discharge 2. Trapping at 22 K CH₂⁺ hν λ>300 nm

benzyl cation   tropylium cation

Scheme 6.8

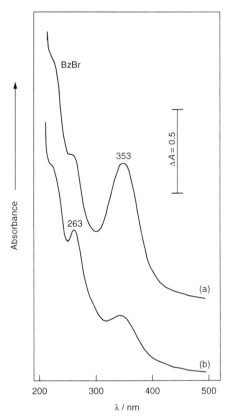

Fig. 6.19 UV–visible spectra of benzyl bromide vapour co-deposited at 22 K with argon subjected to a microwave discharge. (a) After 2 hours deposition; (b) after further photolysis ($\lambda > 290$ nm) for 30 minutes (offset for clarity). BzBr denotes absorption of the precursor. The band with $\lambda_{max} = 353$ nm is assigned to the benzyl cation, while that at 263 nm is assigned to the tropylium cation. *(Adapted with permission from ref. 63. © 1981, American Chemical Society.)*

2.2.8 The naphthalene radical cation

Naphthalene can be photoionized in experiments similar to those described above for benzyl bromide.[64,65] The resulting matrix-isolated radical cation has a beautiful UV–visible absorption spectrum containing a series of well separated vibronic bands at the red end of the visible region. The cation can also be generated by mercury-arc photolysis of Ar matrices containing naphthalene together with CCl_4 as an electron trap.

2.3 Metal carbonyls

A number of metal carbonyls are available commercially, for example $Fe(CO)_5$, $Fe_2(CO)_9$, $Cr(CO)_6$, $Mn_2(CO)_{10}$, and $Mo(CO)_6$. Their matrix photochemistry has provided a rich field for study and a few straightforward experiments selected from this area can be recommended as exercises in matrix isolation. The presence of strong $v(CO)$ bands in the IR spectra of both the starting materials and the matrix products makes the metal carbonyls particularly amenable to study by matrix IR spectroscopy.

2.3.1 Matrix photolysis of Fe(CO)$_5$ to give Fe(CO)$_4$ and Fe(CO)$_3$

Commercial $Fe(CO)_5$ can be purified by trap-to-trap distillation on a vacuum line and is volatile enough to be incorporated with the host in premixed matrix gases. When photolysed in matrices, $Fe(CO)_5$ loses first one carbonyl to give $Fe(CO)_4$, then a second to give $Fe(CO)_3$.[66–68] Figure 6.20 shows IR spectra of $Fe(CO)_5$ in a methane matrix, before and after photolysis. On initial photolysis, the IR bands of $Fe(CO)_5$ diminished in intensity, while bands which can be assigned to $Fe(CO)_4$ appeared (marked 4 in Fig. 6.20(b)). In addition, there appeared two weak bands assigned to $Fe(CO)_3$ (marked 3). On further, photolysis, the bands of $Fe(CO)_4$ diminished, while those of $Fe(CO)_3$ grew (Fig. 6.20(c)).

The identification of the iron carbonyl species involved in these photoprocesses was carried out with the aid of ingenious experiments utilizing ^{13}CO labelling, and the structure of $Fe(CO)_4$ was determined to have C_{2v} symmetry. The reader should refer to the original papers for a full explanation. In the same studies, it was also found that the decarbonylation processes could be reversed to some extent by light of longer wavelengths, and even by radiation from the Nernst glower of the IR spectrometer used. In later investigations, IR-induced isomerization of matrix isolated $Fe(CO)_4$ was discovered to proceed via a novel pseudorotation.[69–71]

Matrix-gas mixtures containing $Fe(CO)_5$ can be prepared easily following Protocol 6, Chapter 4, and deposited as matrices following Protocol 7, Chapter 4. One experiment which can be recommended is an exploration of the matrix photolysis of $Fe(CO)_5$ with UV light and possibly also the partial reversal of the reaction by photolysis with wavelengths greater than 375 nm.

2.3.2 The matrix photolysis of Cr(CO)$_6$ to give Cr(CO)$_5$

$Cr(CO)_6$ is another of the commercially available metal carbonyls. Like $Fe(CO)_5$, it is just volatile enough to be incorporated into premixed matrix gases and offers another convenient exercise in matrix photolysis. The resulting $Cr(CO)_5$ forms complexes with a range of matrix hosts, and although the IR bands of $Cr(CO)_5$ are not very sensitive to this complexation, the electronic absorption in the visible region is sensitive to an unusual extent. The interaction between $Cr(CO)_5$ and matrix host molecules has been discussed briefly in Chapter 1 (Section 2.2.4); the polarized photochemistry of $Cr(CO)_5$ was given as an example in Chapter 5 (Section 2.4.2).

It would be tedious to repeat the matrix photolysis of $Cr(CO)_6$ in a whole range of matrices, simply as an exercise, but straightforward experiments using

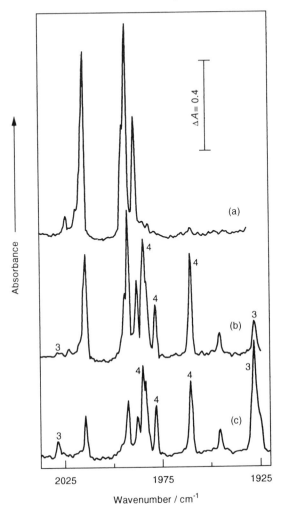

Fig. 6.20 IR spectra, in the carbonyl region, of $Fe(CO)_5$ in a CH_4 matrix at 20 K. (a) Before photolysis; (b) after 1 minute photolysis (medium-pressure Hg arc); (c) after a further 4 minutes photolysis. The matrix ratio $(CH_4:Fe(CO)_5)$ was 10 000:1 and the matrix was deposited by the pulse method. Bands marked 4 are assigned to $Fe(CO)_4$, while those marked 3 are assigned to $Fe(CO)_3$. *(Adapted with permission from ref. 67. © 1974, Royal Society of Chemistry.)*

either IR or UV–visible spectroscopy can be readily selected from the original papers.[72–76] $Cr(CO)_6$ is an excellent case for study with polarized light.

3. Finale

It has been the intention in this chapter to present a selection of experiments that are reasonably easy to carry out and which illustrate as wide a range as possible of the applications of matrix isolation in chemistry. Indeed, it is hoped that the entire book will provide sound advice on the basic techniques necessary for matrix studies, which will be useful for inorganic, organic, and physical chemists alike. Looking back on the whole work, however, the author is conscious of a bias towards his own area of matrix isolation, namely the study of reactive organic species, particularly in photochemical reactions.

This bias has arisen despite the very best intentions to the contrary. But the advantages of first-hand experience, the efficiency of using illustrative material, such as matrix spectra, which is already to hand, and a pardonable preference for knowing what one is talking about, at least most of the time, all exerted a pull in the direction of familiar ground. As a consequence, it will no doubt be judged by many that the author's own work has found its way into this book, as exemplifying matrix research, to a degree unjustified by its scientific merit. It is hoped that the general guidance contained in these pages will be of value, none the less.

The suggested experiments in this chapter cover the spectroscopic techniques of IR, UV–visible, and ESR spectroscopy and the generation of reactive species by photolysis, pyrolysis, chemical reaction, and microwave discharge. The author has virtually no experience of other forms of matrix spectroscopy, or the generation of reactive species by such methods as X-irradiation, electron bombardment, or laser ablation, and apologizes to those of his colleagues active in matrix isolation who may have hoped for some discussion of the practical requirements of these techniques. The same can be said for the production of high-temperature vaporizing species, which commonly require furnace temperatures in excess of 2000 K; and the same apology is made to those working in this field. In mitigation, however, it will be appreciated that no one is likely to construct such expensive items as 3000 K furnaces or platinum Knudsen cells just for a few trials of a new matrix set-up; so the inclusion of experiments requiring such equipment could appeal to only a small minority of readers.

It may also be said that, although the originals of the selected experiments in this chapter span a period of more than three decades, they do not include examples of 'state-of-the-art' research in matrix isolation, requiring specialized and expensive equipment, such as high-power lasers or very high resolution spectrometers. Rather than present these recent developments as the desirable norm, it has been the aim of this book to show how matrix experiments can be carried out with more standard equipment and with a handful of straightforward skills. Matrix chemists, having apparently no less than average common sense, tended to carry out the easy experiments early on in the history of the subject; so quite a few of the examples come from papers published in the 1960s.

The question of the selection of experiments for this chapter reminds the author of another of his numerous anxieties, because he is quite uncertain whether his many friends and colleagues in matrix-isolation research will be more offended to have been left out or to have been included. Some of those whose experiments have been chosen are likely to regard themselves as still too young to be thought of as 'classics' and not take very kindly either to the implied suggestion that their research was, in some way, simple or straightforward. Some of those left out will no doubt supply a host of good reasons why this was an error. The real hope is that there will be enough in this book to introduce matrix isolation to newcomers to the field, and to stimulate novel ideas for exciting research in which matrix isolation can play a part.

References

1. Hallam, H. E. In *Vibrational Spectroscopy of Trapped Species*, ed. H. E. Hallam. Wiley: London, **1973**, Ch. 3, pp. 67–132.
2. Barnes, A. J.; Hallam, H. E.; Scrimshaw, G. F. *Trans. Faraday Soc.* **1969**, *65*, 3150–3158.
3. Barnes, A. J.; Hallam, H. E. *Trans. Faraday Soc.* **1970**, *66*, 1920–1931.
4. Barnes, A. J.; Hallam, H. E. *Trans. Faraday Soc.* **1970**, *66*, 1932–1940.
5. Nelander, B.; Nord, L. *J. Phys. Chem.* **1982**, *86*, 4375–4379.
6. Yeo, G. A.; Ford, T. A. *Spectrochim. Acta, Part A* **1991**, *47*, 485–492.
7. Engdahl, A.; Nelander, B. *J. Chem. Phys.* **1989**, *91*, 6604–6612.
8. Ault, B. S.; Pimentel, G. C. *J. Phys. Chem.* **1973**, *77*, 1649–1653.
9. Ault, B. S.; Steinback, E.; Pimentel, G. C. *J. Phys. Chem.* **1975**, *79*, 615–620.
10. Barnes, A. J.; Kuzniarski, J. N. S.; Mielke, Z. *J. Chem. Soc., Faraday Trans. 2* **1984**, *80*, 465–476.
11. Purnell, C. J.; Barnes, A. J.; Suzuki, S.; Ball, D. F.; Orville-Thomas, W. J. *Chem. Phys.* **1976**, *12*, 77–87.
12. Durig, J. D.; Bush, S. F.; Baglin, F. G. *J. Chem. Phys.* **1968**, *49*, 2106–2117.
13. Goldfarb, T. D.; Khare, B. N. *J. Chem. Phys.* **1967**, *46*, 3379–3388.
14. Mielke, Z.; Ratajczak, H.; Wiewiórowski, M.; Barnes, A. J.; Mitson, S. J. *Spectrochim. Acta, Part A* **1986**, *42*, 63–68.
15. Vegard, L. *Nature* **1924**, *113*, 716–717.
16. Bass, A. M.; Broida, H. P. *Phys. Rev.* **1956**, *101*, 1740–1747.
17. Cole, T.; Harding, J. T.; Pellam, J. R.; Yost, D. M. *J. Chem. Phys.* **1957**, *27*, 593–594.
18. Foner, S. N.; Jen, C. K.; Cochran, E. L.; Bowers, V. A. *J. Chem. Phys.* **1958**, *28*, 351–352.
19. Brown, H. W.; Pimentel, G. C. *J. Chem. Phys.* **1958**, *29*, 883–888.
20. Jacox, M. E.; Rook, F. L. *J. Phys. Chem.* **1982**, *86*, 2899–2904.
21. Jacox, M. E. *J. Phys. Chem.* **1984**, *88*, 3373–3379.
22. Hartung, W. H.; Crossley, F. *Org. Synth.* **1943**, *Coll. Vol. 2*, 363–364.
23. Rook, F. L. *J. Chem. Eng. Data* **1982**, *27*, 72–73.
24. Ghosh, P. N.; Guenthard, H. H. *Spectrochim. Acta, Part A* **1981**, *37*, 1055–1065.
25. Rook, F. L. *J. Mol. Spectrosc.* **1982**, *93*, 101–116.
26. Herzberg, G.; Ramsay, D. A. *Proc. Roy. Soc. (London)* **1955**, *A233*, 34–54.
27. Ewing, G. E.; Thompson, W. E.; Pimentel, G. C. *J. Chem. Phys.* **1960**, *32*, 927–932.
28. Milligan, D. E.; Jacox, M. E. *J. Chem. Phys.* **1964**, *41*, 3032–3036.
29. Ogilvie, J. F. *Spectrochim. Acta, Part A* **1967**, *23*, 737–750.
30. Adrian, F. J.; Cochran, E. L.; Bowers, V. A. *J. Chem. Phys.* **1962**, *36*, 1661–1672.

31. Adrian, F. J.; Bohandy, J.; Kim, B. F. *J. Chem. Phys.* **1984**, *81*, 3805–3810.
32. Andrews, L.; Pimentel, G. C. *J. Chem. Phys.* **1967**, *47*, 3637–3644.
33. Milligan, D. E.; Jacox, M. E. *J. Chem. Phys.* **1967**, *47*, 5146–5156.
34. Snelson, A. *J. Phys. Chem.* **1970**, *74*, 537–544.
35. Andrews, L. *J. Chem. Phys.* **1968**, *48*, 972–979.
36. Andrews, W. L. S.; Pimentel, G. C. *J. Chem. Phys.* **1966**, *44*, 2361–2369.
37. Nelson, L. Y.; Pimentel, G. C. *Inorg. Chem.* **1967**, *6*, 1758–1759.
38. Howard, W. F.; Andrews, L. *J. Am. Chem. Soc.* **1974**, *96*, 7864–7868.
39. Boal, D.; Ozin, G. A. S. *Spectrosc. Lett.* **1971**, *4*, 43–46.
40. Beattie, I. R.; German, A.; Blayden, H. E.; Brumbach, S. B. *J. Chem. Soc., Dalton Trans.* **1975**, 1659–1662.
41. Binnewies, M.; Lakenbrink, M.; Schnöckel, H. *Z. Anorg. Allg. Chem.* **1983**, *497*, 7–12.
42. Hastie, J. W.; Hauge, R. H.; Margrave, J. L. *J. Am. Chem. Soc.* **1969**, *91*, 2536–2538.
43. Maass, G.; Hauge, R. H.; Margrave, J. L. *Z. Anorg. Allg. Chem.* **1972**, *392*, 295–302.
44. Chapman, O. L.; Le Roux, J.-P. *J. Am. Chem. Soc.* **1978**, *100*, 282–285.
45. Donnelly, T.; Dunkin, I. R.; Norwood, D. S. D.; Prentice, A.; Shields, C. J.; Thomson, P. C. P. *J. Chem. Soc., Perkin Trans. 2* **1985**, 307–310.
46. Hayes, J. C.; Sheridan, R. S. *J. Am. Chem. Soc.* **1990**, *112*, 5879–5881.
47. Platz, M. S. *Acc. Chem. Res.* **1995**, *28*, 487–492.
48. Lindsay, R. O.; Allen, C. F. H. *Org. Synth.* **1955**, *Coll. Vol. 3*, 710–711.
49. Dunkin, I. R. *J. Chem. Soc., Chem. Commun.* **1982**, 1348–1350.
50. Dunkin, I. R. *Spectrochim. Acta, Part A.* **1986**, *42*, 649–655.
51. Baird, M. S.; Dunkin, I. R.; Poliakoff, M. *J. Chem. Soc., Chem. Commun.* **1974**, 904–905.
52. Baird, M. S.; Dunkin, I. R.; Hacker, N.; Poliakoff, M.; Turner, J. J. *J. Am. Chem. Soc.* **1981**, *103*, 5190–5195.
53. Weil, T.; Cais, M. *J. Org. Chem.* **1963**, *28*, 2472.
54. Bell, G. A.; Dunkin, I. R. *J. Chem. Soc., Chem. Commun.* **1983**, 1213–1215.
55. Bell, G. A.; Dunkin, I. R.; Shields, C. J. *Spectrochim. Acta, Part A* **1985**, *41*, 1221–1227.
56. Dunkin, I. R.; Shields, C. J. *J. Chem. Soc., Chem. Commun.* **1986**, 154–156.
57. Sander, W. *Angew. Chem., Int. Ed. Engl.* **1990**, *29*, 344-354; *Angew. Chem.* **1990**, *102*, 362–372.
58. Dunkin, I. R.; McCluskey, A. *J. Photochem. Photobiol. A: Chem.* **1993**, *74*, 159–164.
59. Bell, G. A.; Dunkin, I. R. *J. Chem. Soc., Faraday Trans. 2* **1985**, *81*, 725–734.
60. Maier, G.; Lothar, H. F.; Hartan, H.-G.; Lanz, K.; Reisenauer, H. P. *Chem. Ber.* **1985**, *118*, 3196–3204.
61. Angell, C. L.; Hedaya, E.; McLeod, D., Jr. *J. Am. Chem. Soc.* **1967**, *89*, 4214–4216.
62. Kasai, P. H.; McLeod, D., Jr. *J. Am. Chem. Soc.* **1972**, *94*, 7975–7981.
63. Andrews, L.; Keelan, B. W. *J. Am. Chem. Soc.* **1981**, *103*, 99–103.
64. Andrews, L.; Blankenship, T. A. *J. Am. Chem. Soc.* **1981**, *103*, 5977–5979.
65. Andrews, L.; Kelsall, B. J.; Blankenship, T. A. *J. Phys. Chem.* **1982**, *86*, 2916–2926.
66. Poliakoff, M.; Turner, J. J. *J. Chem. Soc., Dalton Trans.* **1973**, 1351–1357.
67. Poliakoff, M. *J. Chem. Soc., Dalton Trans.* **1974**, 210–212.
68. Poliakoff, M. *Chem. Soc. Rev.* **1978**, *7*, 527–540.
69. Davies, B.; McNeish, A.; Poliakoff, M.; Turner, J. J. *J. Am. Chem. Soc.* **1977**, *99*, 7573–7579.

70. Poliakoff, M.; Davies, B.; McNeish, A.; Tranquille, M.; Turner, J. J. *Ber. Bunsenges. Phys. Chem.* **1978**, *82*, 121–124.
71. Poliakoff, M.; Ceulemans, A. *J. Am. Chem. Soc.* **1984**, *106*, 50–54.
72. Graham, M. A.; Poliakoff, M.; Turner, J. J. *J. Chem. Soc. A* **1971**, 2939–2948.
73. Perutz, R. N.; Turner, J. J. *J. Am. Chem. Soc.* **1975**, *97*, 4791–4800.
74. Burdett, J. K.; Perutz, R. N.; Poliakoff, M.; Turner, J. J. *J. Chem. Soc., Chem. Commun.* **1975**, 157–159.
75. Burdett, J. K.; Graham, M. A.; Perutz, R. N.; Poliakoff, M.; Rest, A. J.; Turner, J. J.; Turner, R. F. *J. Am. Chem. Soc.* **1975**, *97*, 4805–4808.
76. Burdett, J. K.; Downs, A. J.; Gaskill, G. P.; Graham, M. A.; Turner, J. J.; Turner, R. F. *Inorg. Chem.* **1978**, *17*, 523–532.

$$\boxed{\text{A1}}$$

List of suppliers

The following list provides contact information for suppliers of items needed in matrix-isolation research, particularly cryostats, vacuum equipment, optical components, and high purity gases. For the most part, the author has had direct experience of the products of the companies included. The list is far from comprehensive. Apologies are due to the numerous other supplier companies, world-wide, that might have been included in a comprehensive list, had it been practicable to devise one. For readers of the book, however, it seems better to provide the assistance of a particular list, as a starting point in the search for equipment and materials, than no list at all. In many countries, local suppliers will be found for the more common items of equipment, such as electrical and vacuum components, either offering their own products or acting as agents for the manufacturers.

Every effort has been made to ensure that addresses, telephone numbers and fax numbers were correct when the book went to press. In this market sector, however, as in many others, companies often change name, merge, re-locate, or go out of business; so details are likely to change over time. Many companies are now developing sites on the World Wide Web, and this is another way of finding contact addresses.

Air Products
UK: Air Products Ltd, Hampshire International Business Park, Crockford Lane, Chineham, Basingstoke, Hampshire RG24 8YP, Tel. 01256-707707; Fax 01256-706326
USA: Air Products and Chemicals Inc., 7201 Hamilton Boulevard, Allentown, Pennsylvania 18195. Tel 1-610-481-4911; Fax 1-610-481-5900

Gases, including high purity gases; gas valves and regulators

AMKO
Germany: AMKO wissenschaftlich-technische Instrumente GmbH, Gärtnerweg, D-25436 Tornesch. Tel. 0412-2510-61; Fax 0412-2549-14
UK: AMKO Scientific Instruments Ltd, 3 Crowley Avenue, Whickham, Newcastle-upon-Tyne NE16 4TD. Tel. 0191-488-6335; Fax 0191-488-6329

Light sources and monochromators

APD Cryogenics

UK: 2 Eros House, Calleva Industrial Park, Aldermaston, Berkshire RG7
8LN. Tel. 01189-819373; Fax 01189-817601
USA: APD Cryogenics Inc., 1833 Vultee Street, Allentown, Pennsylvania
18103. Tel. 1-610-791-6700; Fax 1-610-791-0440

'Displex' closed cycle helium refrigerators, including a 4 K model; vacuum
shrouds for matrix isolation, including a model for ESR spectroscopy

Apollo Scientific Ltd

UK: Unit 1A, Bingswood Industrial Estate, Whalley Bridge, Derbyshire
SK23. Tel. 0161-477-4558; Fax 0161-477-3889

Spectroscopic windows and polishing kits; IR polarizers; a wide range of
accessories for IR and UV-visible spectroscopy

BOC

UK: BOC Ltd, Speciality Gases, The Priestley Centre, The Surrey Research
Park, Guildford, Surrey GU2 5XY. Tel. 0800-020-800; Fax 0800-136-601
USA: BOC Gases, 575 Mountain Avenue, Murray Hill, New Jersey 07974-
2082. Tel. 1-908-464-8100; Fax 1-908-464-9015

High purity gases; gas valves and regulators

Buck Scientific Inc.

USA: 58 Fort Point Street, East Norwalk, Connecticut 06855. Tel 1-203-853-
9444; Fax 1-203-853-0569

Spectroscopic windows and accessories

Coherent-Ealing

France: Domaine Technologique de Saclay, Batiment AZUR, 4 rue René
Razel, 91892 Orsay Cedex. Tel. 16-019-4040; Fax 16-019-4000
Germany: Dieselstrasse 5b, D-64807 Dieburg. Tel. 0607-1968-302; Fax 0607-
1968-499
UK: Coherent-Ealing Europe Ltd, Greycaine Road, Watford WD2 4PW.
Tel. 01923-242261; Fax 01923-244461
USA: 2303 Lindbergh Street, Auburn, California 95602. Tel. 1-530-889-5365;
Fax 1-530-889-5366; 209 West Central Street, Natick, Massachusetts 01760.
Tel. 1-508-655-4330; Fax 1-508-647-1819

UV-visible lamps; optical components, including quartz lenses and filters

Cryogenic Ltd

UK: Acton Park Industrial Estate, The Vale, London W3 7QE. Tel. 0181-
743-6049; Fax 0181-749-5315

Helium cryostats, including ^{3}He and dilution refrigerators

Cryophysics
Switzerland: Cryophysics SA, 39 rue Rothschild, CH-1202 Geneva. Tel. 022-732-9520; Fax 022-738-5246
UK: Cryophysics Ltd., Unit 6, Thorney Leys Business Park, Witney, Oxon OX8 7GE. Tel. 01993-773681; Fax 01993-705826

Offices also in France and Germany, and the Czech Republic

European agents for CTI-Cryogenics closed cycle refrigerators, Janis cryostats, and Lake Shore temperature measurement and control units; custom-built vacuum shrouds

CTI-Cryogenics
UK: Fleming Road, Kirkton Campus, Livingstone, West Lothian EH54 7BN. Tel. 01506-460017; Fax 01506-411122
USA: 9 Hampshire Street, Mansfield, Massachusetts 02048. Tel. 1-508-337-5154; Fax 1-508-337-5464

Closed cycle helium refrigerators

Edwards High Vacuum International
Germany: Edwards Hochvakuum GmbH, Postfach 1409, D-35004 Marburg. Tel. 0642-0824-10; Fax 0642-0824-111
UK: Manor Royal, Crawley, West Sussex RH10 2LW. Tel. 01293-528844; Fax 01293-533453
USA: 1 Edwards Park, 301 Ballardale Street, Wilmington, Massachusetts 01887. Tel. 1-508-658-5410 or 800-848-9800; Fax 1-508-658-7969

Offices also in Belgium, Brazil, Canada, France, Hong Kong, Italy, Japan, Korea, Switzerland, and Singapore

Vacuum pumps and complete vacuum systems, vacuum gauges, valves and other vacuum components and accessories, including flanged stainless-steel flexible pipelines; vacuum pump oils, high vacuum greases and waxes, and Apiezon Q sealing compound; closed cycle helium refrigerators

Farnell Electronic Components Ltd
UK: Canal Road, Leeds LS12 2TU. Tel. 0113-263-6311; Fax 0113-263-3411

Electrical and electronic components, including Peltier-effect heat pumps

Graseby Specac
UK: Graseby Specac Ltd, River House, 97 Cray Avenue, St Mary Cray, Orpington, Kent BR5 4HE. Tel. 01689-873134; Fax 01689-878527
USA: Graseby Specac Inc., 500 Technology Court, Smyrna, Georgia 30082-5211. Tel. 1-770-319-9999 or 800-241-6898; Fax 1-770-319-0336

Agents in many countries

Spectroscopic windows and polishing kits; IR polarizers; a wide range of accessories for IR and UV-visible spectroscopy

Hoke
USA: Hoke Inc., 1 Tenakill Park, Cresskill, New Jersey 07626. Tel. 1-201-568-9100; Fax 1-201-568-5913
UK: Hoke International Ltd, 1–3 Bouverie Road, Harrow, Middlesex HA1 4HB. Tel. 0181-423-0113; Fax 0181-864-7008; Suite 1, The McNeill Business Centre, Greenbank Crescent, East Tullos Industrial Estate, Aberdeen AB12 3BG. Tel. 01224-879222; Fax 01224-879988

Valves for vacuum systems and gas handling

Hymatic Engineering Company Ltd
UK: Burnt Meadow Road, North Moons Moat, Redditch, Worcestershire B98 9HJ. Tel. 01527-64931; Fax 01527-591117

Very compact and lightweight single-stage closed cycle refrigerators, cooling to about 60 K

Janis Research Company
USA: 2 Jewel Drive, PO Box 696, Wilmington, Massachusetts 01887-0696. Tel. 1-508-657-8750; Fax 1-508-658-0349

Cryostats and refrigerators, including closed cycle refrigerators, liquid helium, ^{3}He and dilution systems, liquid nitrogen cryostats, and ancillary equipment

Javac
Australia: Javac (Pty) Ltd, 54 Rushdale Street, Knoxfield, Victoria 3180. Tel. 03976-37633; Fax 03976-32756
UK: Javac (UK) Ltd, Unit 6, Drake Court, Brittania Park, Middlesbrough, Cleveland TS2 1RS. Tel. 01642-232880; Fax 01642-232870

Rotary vacuum pumps, vacuum gauges, valves and other vacuum components and accessories, flanged stainless-steel flexible pipelines, vacuum pump oils

Jencons Scientific Ltd
UK: Cherrycourt Way Industrial Estate, Stanbridge Road, Leighton Buzzard, Bedfordshire LU7 8UA. Tel: 01525-372010; Fax 01525-379547
USA: Bursca Business Park, 800 Bursca Drive, Suite 801, Bridgeville, Pennsylvania 15017. Tel. 1-412-257-8861; Fax 1-412-257-8809

Quartz and glass windows

Kurt J. Lesker Company
USA: 1515 Worthington Avenue, Clairton, Pennsylvania 15025. Tel. 1-412-233-4200 or 800-245-1656; Fax 1-412-233-4375

UK: Kurt J. Lesker Company Ltd, 16 Ivyhouse Lane, Hastings, East Sussex TN35 4NN. Tel 01424-719101; Fax 01424-421160

Offices also in Hungary, and agents in Israel, Italy, Korea, Portugal, and Spain

Vacuum systems and accessories, including custom-built vacuum chambers

Lake Shore Cryotronics Inc.
UK: 64 East Walnut Street, Westerville, Ohio 43081-2399. Tel. 1-614-891-2243; Fax 1-614-891-1392

Temperature measurement and control: silicon diodes and thermocouple sensors and electronic control units; speciality accessories such as cryogenic wire, indium foil, thermally conducting grease, electrical feedthroughs, and cryogen transfer lines

Leybold
Germany: Leybold Vakuum GmbH, Bonner Strasse 498, D-50968, Köln. Tel. 0221-347-0; Fax 0221-347-1250
UK: Leybold Ltd, Waterside Way, Plough Lane, London SW17 7AB. Tel. 0181-971-7000; Fax 0181-971-7001

Vacuum pumps, systems and accessories; closed cycle refrigerators, including a 4 K model

Lightpath Optical Company Ltd
UK: Unit 3, The Elms Industrial Estate, Church Road, Harold Wood, Essex RM3 0JU. Tel. 01708-349136; Fax 01708-381638

Accessories for IR and UV-visible spectroscopy

Melles Griot
Germany: Melles Griot GmbH, Lilienthalstrasse 30-32, D-64625 Bensheim. Tel. 0625-1840-60; Fax 0625-1840-622
The Netherlands: Melles Griot BV, Hengelder 23, PO Box 272, 6900 AG Zenvenaar. Tel. 0316-333041; Fax 0316-528187
UK: Melles Griot Ltd, 2 Pembroke Avenue, Waterbeach, Cambridge CB5 9QR. Tel. 01223-203300; Fax 01223-203311
USA: 55 Science Parkway, Rochester, New York 14620. Tel. 1-716-244-7220; Fax 1-716-244-6292; US offices also in Auburn MA, Boulder CO, Carlsbad CA, and Irvine CA

Offices also in Canada, Denmark, France, Israel, Japan, Singapore, and Sweden; agents in Australia, China, Italy, Korea, Spain, Taiwan, and Switzerland

Optical components including fused silica lenses, sapphire windows, filters, and UV-visible and IR polarizers

Nicolet Instruments Ltd
UK: Budbrook Road, Warwick CV34 5XH. Tel. 01926-417700; Fax 01926-417701

Accessories for IR spectroscopy

Oriel Corporation
USA: 250 Long Beach Boulevard, Stratford, Connecticut 06497-0872. Tel. 1-203-377-8282. Fax 1-230-378-2457
UK: L.O.T. – Oriel Ltd, 1 Mole Business Park, Leatherhead, Surrey KT22 7AU. Tel. 01372-378822; Fax 01372-375353

Offices in France, Germany, Italy and Switzerland, and agents in many other countries

UV-visible lamps and optics, including filters, monochromators, and polarizers

Oxford Instruments
UK: Research Instruments, Tubney Woods, Abingdon OX13 5QX. Tel. 01865-393200; Fax 01865-393333
USA: 130A Baker Avenue, Concord, Massachusetts 01742. Tel. 1-978-369-9933; Fax 1-978-369-6616

Offices also in Australia, China, France, Germany, Japan, and Singapore

Liquid helium, liquid nitrogen, and closed cycle helium cryostats; temperature controllers and accessories; systems for IR, UV-visible, Raman, ESR, and Mössbauer spectroscopy

RS Components Ltd
UK: Birchington Road, Weldon Industrial Estate North, Corby, Northamptonshire NN17 9RS. Tel. 01536-201234; Fax 01536-405678

Offices in Australia, Austria, Denmark, France, Germany, India, Italy, Republic of Ireland, and New Zealand

General mechanical, electrical, and electronic components, including Peltier-effect heat pumps

Spectra-Tech
USA: Spectra-Tech Inc., 652 Glenbrook Road, Stamford, Connecticut 06906. Tel. 1-203-357-7055 or 800-243-9186; Fax 1-203-357-0609

Accessories for IR spectroscopy

Speirs Robertson
UK: Speirs Robertson and Co. Ltd, Oakley Road, Bromham, Bedford MK43 8HY. Tel. 01234-344775; Fax 01234-825819

USA: 5625 FM, 1960 W #610, Houston, Texas 77069. Tel. 800-747-4596

UV-visible lamps and optical components, including quartz lenses, filters, monochromators, and UV-visible polarizers

J. Young (Scientific Glassware) Ltd
UK: 11 Colville Road, Acton, London W3 8BS. Tel. 0181-992-0891; Fax 0181-992-0891

Agents in the USA

Glass vacuum taps and couplings; gas bulbs

Index